James Hunter Crawford

Wild life of Scotland

James Hunter Crawford

Wild life of Scotland

ISBN/EAN: 9783743341869

Manufactured in Europe, USA, Canada, Australia, Japa

Cover: Foto ©berggeist007 / pixelio.de

Manufactured and distributed by brebook publishing software (www.brebook.com)

James Hunter Crawford

Wild life of Scotland

Wild Life of Scotland

By J. H. CRAWFORD, F.L.S.

ILLUSTRATIONS BY JOHN WILLIAMSON

LONDON: JOHN MACQUEEN

MDCCCXCVI

CONTENTS

1

INTRODUCTION

I N times less secure than our own, refuges were placed down, at intervals, over the untamed country, according to the special needs and dangers of the place. A common site was the river bank, or the entrance to the mountain pass, where the traveller might rest till the brown water subsided, or the daylight returned.

Scotland abounds in flooded streams, and stern passes. At the southern end of a road, leading over the central Grampian ridge to and from Braemar, stands the 'spital, or hospital of Glenshee. Everything seems to point to the genesis of the present hamlet in an ancient shelter. The scene is typical. And if, in imagination, we refill the shades, cast by the then dense woodland, with wild beasts, the reason for the choice will become still more apparent. It must be difficult for those who now

halt by the way at the inn, to realise a past state of
things.

As a last resource, much of the wood was destroyed;
and the tenants thereof dealt with on more equal
terms. So began a war of extermination, only
justifiable as a measure of self-preservation, which
was not without its sacrifices. The mischief, as far
as the trees are concerned, we are slowly repairing;
often with no other view than to soften and vary the
hardness and monotony of the landscapes. What
about the animals?

Toward the close of last century, a temporary
movement began; which soon spent itself, and
seems to have excited so little attention, that,
probably, its very existence is not generally
known.

Its object was to restore some of the exiles
thus, unceremoniously, driven forth. The boldest
advocate of things as they were, acknowledged that
these islands were too contracted for the larger, and
more dangerous of the wild animals to be turned
loose; and drew the line at the bear, and the
wolf.

But, it seemed to some that it might increase the
picturesqueness even of such hills as ours, if the
reindeer were once more to walk along the ridges,

in relief against the evening sky; and that it might add fresh interest to our streams if the beaver, known to our forefathers as the water-dog, were to restore his ancient lodges; and show how to build a dam, that would resist the most impetuous rush of our winter spates.

A very few voices were raised in favour of the wild boar; whose presence would revive, it was thought, the fear and wonder of the woods. To those who have the courage to do and to dare, and no one else has a right to handle gun or spear, the game is none the worse of having a tusk, and the will to use it.

The great-uncle of the present Duke of Fife turned some wild boars into the forest of Mar; but the experiment failed, for lack of oaks to shed their acorns, and meres to grow their water-lilies.

Obviously, the conditions of life must be restored before the living creatures. If still larger areas were forested, which ere long will probably be the case, the difficulty will cease to exist. Meantime, some of the scanty patches, which so shabbily represent our ancient woodlands, might shelter a few. What those restorers desired was not a country overrun, as of yore, either with trees, or wild animals; but such a number of boars, in suitable places, sufficiently far out of the way of the nervous, as would place the name once again on the list of our fauna.

"About 1796," writes the Duke of Atholl, " my grandfather imported several reindeer from Lapland, but they nearly all died on the journey between Leith and Blair - Atholl, I believe for want of some particular moss on which they feed. One, however, I understand, lived for two years. I see no reason why reindeer should not be introduced into this country, looking to the more rapid means of transit we now possess."

The climate cannot have changed so much since the twelfth century, as to be fatal. There is good reason to believe that the reindeer haunts the region of the glaciers, and the margin of the icefield, less from choice than persecution. A male and female let loose, amid the misty and snowless mounds of Orkney,—conditions altogether different from those with which we associate them,—seem to have perished from neglect; 'not being found to answer the purpose intended.'

The lowly diet, of which there is abundance on the Highland hills, takes little away from that of any other living creature. And, in addition to its personal picturesqueness, the reindeer would, probably, be found of some economic importance.

In 1874–75, the Marquis of Bute placed some beavers on his Mount Stuart estate, where the stream runs through a wood. In 1878, the keeper was sure of sixteen being alive, which made an average increase of four for each season. The last of them died about five years ago, because, it is assumed, there were no more trees to cut down.

With the latest survivor of the restored exiles, the amiable experiment terminated; and, so far as I know, there is no boar, reindeer, or beaver, in a wild state, in all the land.

A decreasing number of herds of wild cattle, with black ears, thus differing from the red-eared Chillingham breed, are still preserved in some of our parks. These, which have been supposed to represent the original wild ox of the country, are more probably related to the tame cattle brought over by our Saxon forefathers. Even so, their preservation, in undiminished purity of colour and descent, is interesting.

Colquhoun was surprised to come across a herd, grazing, tranquilly, by a moorland roadside, in

Argyllshire.   The neat, well-set-on horns, black
muzzle,   snowy   hide,   and   clean - made   limbs,
guaranteed both the antiquity and the purity of
the breed.   Doubtless, this little group would now
be sought for in vain.

It chanced, that in the very scenes where the
exiles reappeared for a brief while,—the forests of
Perthshire and Aberdeen,—and not above a decade
from the death of the last reindeer—say 1810—
another was added to the existing motives for
destruction; and, soon after, a havoc began, which
threatens, not only permanently to unsettle the
balance of life, but even to defeat the very end
in view.   Never, perhaps, was there a more
striking instance of short-sightedness.

Such is the present somewhat transition state
of the wild life of Scotland.   What the future may
be depends a good deal on our action now.

There will be differences of opinion as to the
advisability of remitting the banishment of certain
children of the wilds.   While, among my corre-
spondents, some are in favour of one form, to the
exclusion of the rest; others shake their heads over
the whole matter.   "We cannot afford to grow
wood for beavers to gnaw, or for boars to whet
their tusks on,"

Even to those of us who have been nurtured under present conditions,—for, after all, our natural education only begins with our own lifetime,—the apparition of a reindeer on the hills, or a water-dog in the streams might appear a foreign element; and make Scotland look scarcely like herself. But all ought to be agreed as to the undesirableness of driving away any more, or persecuting those that are left.

Meantime, my scheme is much wider; and will embrace creatures outside the disturbed areas, whose lives are passed under ordinary conditions, and exposed only to natural enemies.

I am touched with a passion for wild nature,— the wilder the better; and confess to a special interest in whatever lives beyond enclosures, and has not been spoiled by that form of taming known as preserving. What I have to say has, at least, so much merit as belongs to personal observations, made, not once, but many times. It has been my delight to visit my friends in their homes, and to go back again and again to see them.

When I move abroad, I carry no stock of pre-possessions. I have been careful to put in practice a lesson I first learned with regard to my own kind, to forget all I had heard, so that I might look and judge for myself.

If, in anything, I have seen wrongly; and anyone who has learned in the same way, will put me right; so far from being annoyed, I shall be glad to learn from wiser men. Often, in the July sun, have I sat down on the stone dyke, shaded by a mountain ash, to listen to a stonebreaker; and I am not likely to complain of information from other quarters.

It is somewhat difficult to deal with a mass of material, and a wide area of country; but my plan is simple. I have chosen out representative experiences; dropped down, here and there, on likely places. And, perhaps, the following chapters will give a fairly complete account of the forms of life in the wilds of Scotland; and in the waters which surge twice a day round her coasts.

WILD LIFE OF SCOTLAND

I

SPRING BIRD LIFE

SPRING awakens under the eaves, whence, in the dark mornings, comes a twittering, like the stirrings of a household from sleep. A knowing and familiar whistle tells that the starlings are back again; and a succession of guttural sounds, repeated as fast as they are able, seem to say that they are just about to begin. The starling, being a mimic by nature, it is easy to assign many of his notes to their proper owner.

When it is just getting light at seven o'clock, the voice of the missel-thrush breaks in upon our waking sense, with the pleasant tale that the season of waiting is past. He is called storm-cock, because the song fills up the intervals of the early spring gales. Less varied, and more interrupted than that of the mavis; less melodious than that of the blackbird, the song is still sufficiently thrush-like to be easily recognised as belonging to that musical family, and sufficiently early to be welcome, and delightful.

While, in England this is a familiar garden bird, building carelessly in bushes and shrubbery; in Scotland he is one of our wildest wood birds, choosing, with unerring instinct, the most difficult trees to climb. The last nest I saw was on an oak sapling, drawn out to portentous length and thinness, in its endeavour to reach the upper light and air; and with a nasty elbow about the middle, round which it was well-nigh impossible to wriggle. I could not help thinking that the builder saw the strategic advantage of that bend in his contest of cunning with the schoolboy.

There is nothing more marked in the bird life of Scotland than the rapid increase of this form. Already, there are signs that he is overflowing from the wilds into the neighbourhood of houses; and, in some places, his song comes as often from the topmost twig of the apple-tree, as from the topmost whorl of the spruce-fir.

A fortnight later, the song-thrush, or mavis·

awakens, and delivers his lay,—clear, sweet, jubilant, continuous, — among the lower branches. Single notes, double notes, notes in strings of three and four are poured forth with a gush, and verve which thrill the listener. Nor does this represent the variety, much less the music of a song, in which spring itself seems to be finding utterance.

Yet another thrush follows on, adding his rich contralto to the mavis' glad treble. He who has passed along the lanes, and by the wood copses, after a shower has cooled the air, freshened the young leaves, and washed the May-flower into a purer white; and listened to the outburst of blackbird music, has been present at one of the most delightful of nature's concerts.

The mavis sings the moment he falls in love. The blackbird, more practical, waits till he sees the first egg in the nest. When satisfied that there can be no mistake, he trolls out a few careless notes, and finishes up with a lazy twitter.

The mavis is a morning bird. His song is a glad matin, breathing the hopefulness of daybreak. The blackbird belongs to the evening, as his very colour would suggest. His song is a vesper, according with the soberness of twilight. His

chirpy scream is the last sound heard, before silence
falls over the plains, and the trees of the wood go
to sleep.

Meantime the chaffinch, or shilfa has been calling
"finch, finch," as it has been rendered; from
which, the whole group to which he belongs has
got the name of finches.

But, much more musical is the Scots "dreep,
dreep, weet, weet"; and with much more meaning,
seeing that it is uttered at the showering, and
growing time of the year, when the soft rain wets
the leaves, "weet, weet"; and, gathering into drops,
drips down on the blue hyacinths, and pink and
white anemones, "dreep, dreep."

These calls of birds seem to belong to the woodlands, where mates are apt to lose each other in
the labyrinth of branches. They are something
apart from the song, and have quite another
mission. They are meant not to please, but to
summon; or to tell where the caller is when he is
wanted. "I am in the beech-tree," they seem to
say; or "I have changed my position to the elm."
Nothing strikes observers in treeless landscapes
more than their absence.

In some such order, appear our resident songsters.
They overlap one another, of course. By the middle

of March, the chaffinch has already begun; the
mavis prolongs a song, which has lost a little of its
gush, into July. In May all sing together. The
only bird that has the world all to itself for a brief
interval, is the missel-thrush. But, there is always
a dominant note; a songster which gives a character
to the month.

The missel-thrush for March, the mavis for April,
the blackbird for May, the chaffinch for June. Or,
if we borrow a charming calendar from the flowers
springing beneath: the missel-thrush with the
celandine, the song-thrush with the anemone, the
blackbird with the primrose and the violet; the
shilfa with the wild hyacinth of the shades, the
May of the hedges, and the cuckoo-flower of the
meadows.

The long-drawn-out call of the late-nesting
greenfinch comes from every leafing tree; and
his musical trill, and "chum, chum, chum, chum"
is the undertone of the woods; inaudible to the
inattentive, but ever present to those who have
their senses open. With the doubtful exception
of the chaffinch, the greenfinch is our commonest
bird.

The colours of spring are quite as charming
as the songs. The russet becomes redder in the

breast, and the blue bluer on the head of the
chaffinch; the green flashes out more vividly in
the male greenfinch; the metallic colours scintillate
more brilliantly on the coat of the starling; and
the beak of the blackbird shows signs of fresh
dipping. But no surprise is greater, or more delightful than the change in the yellow-hammer;
and, nowhere does yellow look so bright and
pleasant in springtime, as in the head of that bird,
—except perhaps it may be on the breast of the
grey wagtail.

Men-children should never lose their passion for
bird-nesting; not with a view of robbing, or even of
disturbing the builder, which would be unmanly;
but, only, to enter once a year into fairyland.

I should as soon think of letting spring pass
without seeing the larch hang out its tassels
against the dark green of the pinewood; the beech
shake out afresh its pale green leaves; the apple-
tree cover itself with blossoms; the cuckoo-flower
blow on the meadows; the veronica reflect the
tender azure of the sky; the forget-me-not reveal
its deeper blue through the grass by the stream-
side; as miss the annual peep, behind the leaves,
and through the thorns, at the mossy nest with its
four oval eggs.

Consider the nest of the long-tailed tit, with its shining adornment of lichens, stitched together with fragile threads, woven in the spider's loom. The lining consists of no less than two thousand feathers. How many farmyards, and rustic lanes, and shaded woodlands, and cool stream-sides has the little bird visited in making this collection? Why, half the ornithology of the district is represented there!

The nests are hard to find; not because they are hidden away where the twigs and branches are thickest, but, because the fabric blends so delicately and tenderly with its surroundings, that it deceives the eye. Though straight before me, without the shelter of a single twig, it is a long time before I can separate each one from its surroundings. The bunch of lichens on the fork of the fir-tree, exactly resembling those which cover the rest of the trunk, is that daintiest sample of bird architecture the nest of the chaffinch.

The female is so soberly, and harmoniously coloured, that I can scarcely tell that she is sitting. Thus, during the critical period, when she is almost constantly on the nest, and an easy prey to any enemy, she has a maximum chance of escaping observation.

The male chaffinch is the brightest of woodland

2

birds. Free to move about, and elude observation, he is safe to retain and even increase his ordinary attractions. Were the sexes once alike, and are we to assume that their respective shares in this interesting duty made all the difference? Many fingers seem to point in this direction.

In the case of ground birds that build on the exposed heather, like the grouse, or in the open field, like the partridge, or on the bushless moor, like the lark, where all are exposed to the same risk, the male is toned down to the harmonious shades of the female. A friend informs me of a case, which came under his own observation, where a hen blackbird which had passed the age for sitting, put on the inky hue of the cock. Probably, a little closer attention would reveal this tendency in other species.

When the hen chaffinch leaves the nest, the wine-coloured eggs are found to blend with the heavy shades of the fir needles. White, or only slightly spotted eggs are seldom exposed in an open nest like this; but are domed over as in the case of the long-tailed tit, or hidden away in holes, like those of the swift, and the sand-martin.

The young birds, so long as they remain in the nest, are still sober coloured; and soon learn to distinguish between the wing of the returning parent bird, and the foot of an intruder, so as to cower down, and keep their yellow beaks closed in time of danger.

Thus bird, nest, and egg are all, more or less, protectively coloured.

Spring is the pairing season, and the courting of the birds, which precedes, or accompanies this process, is a pleasant phenomenon, which seems to be well-nigh universal. This is a more conscious and direct appeal than either colour, or song. Whether vanity is at the root of it, we should require a much more intimate acquaintance with bird psychology to determine.

The rooks in the stubble field gyrate round other rooks, which, persumably, are of the opposite sex, advancing, and retreating, and displaying to the best advantage, attractions, whose chief shortcoming is their monotony. Even the common sparrows seek to impress the sober-coloured hen by spreading out the flight feathers, and dragging the wing along the flags or the Macadam.

Still more demonstrative are the plovers, and their kindred. The male woodcock performs in the air, until he is satisfied that he has duly impressed the clucking hen beneath. And, few lovers of wild nature are unacquainted with the spring rush, or rumble of the snipe's wing, as he descends through the air.

But these are only fragments. In the case of the game-birds, we have an amorous ritual approaching completeness, which is annually re-hearsed from beginning to end. Grouse pair early, and their becking season comes long before the tardy snow is off the heather. They begin soon in the day, as well as in the year. The male bestirs himself to sport with the female in the dim light of the breaking day; long before anyone, but the poacher, is there to see. Rising in the air, he indulges in short playful flights, uttering, as he descends towards some favourite knoll, the peculiar cry which has given rise to the name of "becking." The female responds; and he continues his advances until daylight puts a stop to the merry madness.

More imposing is the ritual, in the case of those species which claim a plurality of wives. The blackcock chooses some open in the woods, and, ere the first flush of dawn, is in his place, and ready to begin. He trails his wings, erects his tail-feathers, makes extraordinary leaps, and fights with rivals. The tournament continues till the victor is left in peaceful possession of his harem.

Our resident birds have progressed in the tender labours of incubation; some of them have raised a brood, before the strangers arrive.

Scarce second in interest to the first note of the missel-thrush, is the appearance of the first migrant. And, among the migrants, none are so welcome, or so characteristic of the beginning of the pleasant months as the warblers.

This group owes its popular name to the nature of the song; which may be loosely defined as all melody, and no tune. In some cases, there is no particular reason, except want of breath, why it should break off at all, so untrammelled, and sweetly wild, and natural is it.

The males generally arrive before the females, and during this period of enforced bachelorhood, they express their impatience in their loudest, and sweetest song. The bird-catchers (if I am informed aright) like to secure them in this interesting interval. The noisiness, and assertiveness of the widowed whitethroats, when I visited their haunts, for the first time, on the third of May, was particularly noticeable.

The chiff-chaff is the earliest visitor. His "chilp, chilp" is heard in the wood, when yet the branches are bare: and the daffodils, if there were any, would "fill the winds of March with beauty." He is said to be a rare bird in Scotland, but I do not find him so; and, it strikes me to suggest that naturalists should add a little innocent bird nesting to their other modes of observation, such as watching, and listening. Or, if they would only stoop to consult that most sensitive of registers (which I hope, without much faith, is becoming a thing of the past), the schoolboy's string of eggs, they would find that the scarcity is only apparent. In a little strip of young wood, I can always rely on finding several nests.

The chiff-chaff is the tiniest of a charming little group of small warblers, familiarly known as wrens, from their habit of building domed nests on, or near the ground. The others are the wood-wren, and the willow-wren. They are classed as leaf searchers, because, though nesting on the ground, they live, and feed in the trees. The prevailing tint of their coat is green, so that, when at work, no enemy can see them. In other words, they are protectively coloured.

If the chiff-chaff is supposed to be the rarest, the willow-wren is, undoubtedly, the most numerous of the warblers. Any strip of wood will yield several; and any single tree, not too far up the hillside, will be almost sure to enclose one in April. Throughout June, his song is, perhaps, the commonest of woodland, and even wayside sounds. His only rival is the chaffinch; and one would need to count, in order to make quite sure which has the pre-eminence. The northern finch, and the southern warbler thus meet on the same branch, and seem to be engaged in a friendly contest. For the first few notes, it is not easy even for the practised ear to tell which is singing; and then the chaffinch waxes louder, and ends in the characteristic vigorous flourish; the effect of which becomes, after a few repetitions, distinctly monotonous.

The willow-wren's is not a song. It has neither tune, nor character of any kind. One must be contented to recognise; but can neither define, nor reproduce it. There is nothing to lay hold on; only an impression. It is disembodied melody. Many a time have I leaned against the trunk, and tried to fix it in some words or symbols, which would recall it when it had passed; and I have always failed. It rises from silence in a few sharp notes, which open out; and dies into silence again, so that one cannot say, there it ended. It is the liquid flow, the inexpressible sweetness, that pleases; and of which one never tires, however often it is repeated.

Comparisons are odious, but, sometimes, one prefers it to that of the thrush. Each has its place, probably has been coloured and attuned through countless ages, by its surroundings. The thrush for the morning, the blackbird for the evening, the willow-wren to give expression to the sweet sadness of the breathless, shadow-flecked summer wood.

In her nest, on the cool bank of the burn, protected by the faded leaves of a fallen beech twig, or overhung by white wood-sorrel, or blue veronica, the hen lays her eight or ten delicately-spotted eggs. And, there she sits, through fourteen summer days, looking out into the vista of branches, and shadows.

The whitethroat, known by the ray of sunlight high up on the neck, is one of a group, very much alike in structure, and habits, which includes the blackcap, and the garden-warbler. He has a harsh chiding note for the passer-by, and a warble, always vigorous, and mainly sweet, for his mate.

The sedge-warbler is the most familiar of an aquatic group, found by stream-sides, and marshy places, of which the only other Scots example is the reed-warbler. His chiding is harsher than the whitethroat's. It is amusing to pass between the brambles on the bank, and the willow or alder bushes by the stream, and listen to the scolding on either side. The song is continuous, and largely imitative, with some notes that remind us of the mavis, and some of the clearer quality of the robin, with a grating accompaniment throughout, which is all his own.

The nightingale deserves a division to himself, and serves as an example that, in bird as in woman, plainness of appearance and beauty of song or spirit may go together. The nightingale is looked for in the lanes, and thickets on May day ; and greets the English villagers around the Maypole. The headquarters are in Surrey, where the bird is said to sing sweeter than elsewhere. The range is from the Isle of Wight to Yorkshire. He is, thus, an English bird, and his relation to Scotland is that only of an absentee. Any report of his appearance here must be received with caution, if not with suspicion. Although, if less persecuted, he might soon become Scots.

An interesting, if unsuccessful, attempt was once made to enlighten the dreary Caithness landscape with nightingale music. Sir John

Sinclair placed eggs, gathered in the neighbourhood of London, in nests built near John o' Groat's
House. The robin was chosen as a foster-parent,
because he is a soft-billed bird like the nightingale,
and would be likely to provide the nestlings with
suitable food. All went well for a while. The
young were safely hatched and reared, and were
frequently observed during the remaining summer
weeks flitting about the bushes. But, in due season
the instinct to migrate stirred within; they faced
southward, and never returned.

It was an act akin to that of a friend, who built
nests of twigs among his beech-trees to induce the
passing rooks to settle. Nature refuses to be aided
or coaxed, except in certain cases. Game, or seed
birds, had they found the necessary conditions of
their life, would have raised no objection. Both
would have spread. But nightingales are only with
us for part of the year; and the gap was too wide
to be crossed all at once.

Far nearer the purpose, if the London bird-
fanciers let the eggs alone, so that the birds might
spread of themselves, overlapping their former
limits season by season. And, who knows, but that
one day the good people of Caithness might waken,
or go to sleep to the sound of nightingale music.

That everything comes to him who waits, is a truth that cannot be too carefully laid to heart, in all our dealings with natural truth.

Up to York, the list of our migrant warblers remains unbroken. There, or thereabout, the nightingale drops behind. The rest cross the Tweed for varying distances, thinning out more and more as they pass farther north. The stronghold of the blackcap seems to be the neighbourhood of Durham, and it remains fairly numerous the length of the Moray Firth, on the east coast. The range of the garden - warbler, according to present reports, is much the same. Both are found sparingly, the Duke of Argyll informs me, in the neighbourhood of Inveraray.

Certain of the warblers are so much like others, or are so retiring in their habits, that it is more likely that they are confounded and overlooked, than that they are absent. The common white-throat alone may be reported, when both species are present. The grasshopper - warbler is the modestest of birds, keeping himself out of sight, and deceiving all but the initiated with his insect-like chirp; and the reed-warbler haunts the tall marsh grass, which converts the swampy islets of our great streams into dense, and impenetrable jungles. The last may be, pretty confidently, looked for amid such favourable surroundings.

Chiff-chaff, lesser whitethroat, and grasshopper-warbler are reported as far north as the Moray Firth.

Three forms, the sedge-warbler, the willow-warbler, and the redstart (in so far as the last can be called a warbler), occur throughout Scotland. Certain of these may be more or less localised by their habits, or may be absent from some districts, and dominant in others for no very apparent reason; but, with these limitations, their distribution is universal. The modest demand, in one case, is simply that a single tree shall break the bareness of the landscape, to yield a branch to sing from, and a wilderness of leaves to search for food. A fourth may, safely, be added to the list of universally distributed species. Scotland's bramble brakes and broomy knowes will yield the white-throat; its lanes, especially in the neighbourhood of houses, the redstart; its marshes and stream-sides, the sedge-warbler; its copses, the willow-wren.

Whatever northern counties some of our Scots warblers may, or may not pass over, all, with the very doubtful exception of the grasshopper-warbler, reappear on the Shetland Isles.

Quite as typical migrants, and perhaps more

generally known are the members of the swallow tribe. The warblers, whose food is mainly in the caterpillar, and soft intermediate stages, may, and occasionally do weather a mild winter; all feeders on insects in the winged state must go, or perish. The lingering of a late brood of swallows has led to the application of a false analogy to account for their disappearance. No bird does hibernate; and, it would be a strange fact if the warmest blooded of all animals did. This year boys were chasing swallows, which could by no means leave the country, in the streets of St. Andrews, as late as the twelfth of November.

The earliest to arrive,—at least they are due on the same day as the chimney-swallows, are the sand-martins. They are the smallest of all known species, and are singular, among our visitors, for their mouse-coloured back, and jerky flight. Last year, they reached their village, in the soft sandstone cliffs of St. Andrews, on the eleventh of April. Next day, the wind changed to the east, bringing the usual chill, and they disappeared; returning on the fifteenth of the same month. Plainly, they had departed south again, beyond the limit of the cold snap, which had probably checked the development of insect life.

The first chimney-swallow for the year was sitting on a branch. He was easily known by his chestnut throat, and the two long feathers, which give such a deep fork to his tail. Most of the swallows do perch and light; but only awkwardly, and as if they were not meant for it. The skimming of this bird over the ground, or the surface of water is singularly swift, and graceful. But for its structure, and habits, it might be classed among the warblers.

The other evening the air seemed full of sweet sounds, which died in the distance, and swelled again. My companion looked to the hedge, and to the tree, and everywhere, but into the air, and it came upon him quite as a pleasant revelation that any swallow could sing. But this is the only songster: the sand-martin is mute, the house-martin simply twitters, the swift screams.

The chimney-swallow builds, by preference, in the rafters. An old mill, with the liberty of the dam for an evening skim, when the insects are near the surface, is an ideal site. But he builds also under the eaves; and even, on occasion, in the window corners.

The house-martin deserts his favourite window corner for the eaves, and occasionally finds his way among the rafters. Thus the domains overlap; and, as is so frequently the case where kindred species are thrown thus closely together, individuals fall in love, and the chocolate-throated swallow mates, and nests with the white-throated martin.

The four species seem to be universally distributed; with a predominance of one over the other, according to local conditions. It is easy to see why it should be so. There are no limitations such as confine the birds of the earth to the hedgerow, or the woodland, or the moor; no barrier in the air to prevent insects spreading, and being as numerous on one side of a fence, or a stream, or a mountain chain, as on the other. All are common on the Border; all are common in Sutherland. From Bute the report is: "The swallow tribe seem to revel in the neighbour-

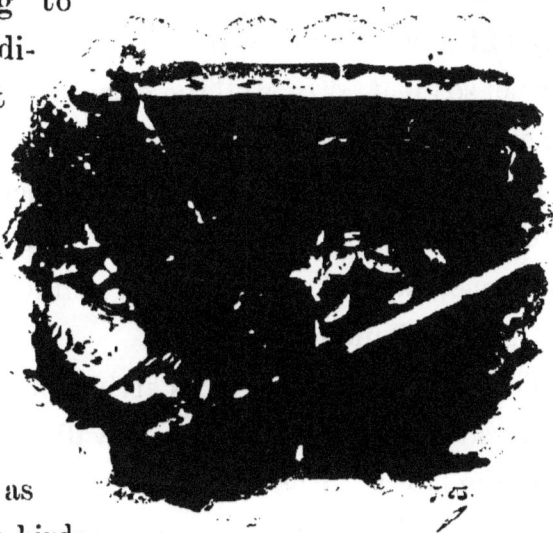

hood." In Shetland the swift is only occasional, the others rare.

The fern-owl is the migrant of the dusk, the swallow of the moth. He occupies the interval between daylight and dark, along with the bat; and, from his habit of circling round the trees, where insect life is most abundant, gets the name of wheel-bird. His distribution is as universal as any of the species of swallow, although the numbers are not so great. "In Mull the monotonous spinning-wheel note was raised each July evening close to both our shooting quarters. In Sutherland he is oftener seen than heard." In Shetland he is commoner than the swallows, probably because he is a ground-builder.

The sea-swallows, or terns arrive about the same time as the earlier of the land-swallows, and leave with the latest. I have a note of 4th October last year, "Swallows and terns still here"; and probably they lingered about for a week, or ten days later. Agassiz makes the Arctic tern scud south, just ahead of the first snow-laden northern gale. In spring, I frequently notice the tern and the chimney-swallows, for the first time, on the same day.

Sea and land swallows are alike in so many particulars that little ingenuity was needed in

classing them under the same popular name. Both
have the same crescent wings, the tern's being sharply
crescent like the swift's, and so long, when expanded,
that they seem almost out of proportion with the
body; both have the same forked tail, the tern's
being deeply forked like the chimney-swallow's;
both have the same swift flight, the tern's flight
being jerky like the sand-martin's.

The migrant of the burn is the common sandpiper, or summer-snipe, which joins the nesting-
dipper early in May, startling the abstracted
angler by its scream, and lending a touch of wildness to an out-of-the-world scene.

The dotterel reaches Scotland in the first, or
second week of May, in small flocks of from six
to a dozen, known as "trips." After a few days'
rest on some coast moor, they scatter thinly over
the uplands. It is said that never more than one
pair is found on the same hilltop; but, probably, this
is only a strong way of stating the undoubted fact
that nowhere are they very numerous. I find
that bird and nest are unknown to all except game-
keepers and shepherds, to whom, however, both are
familiar. This is the migrant of the mountains.

The migrant of cultivation is the corncrake.
The lapwing has already raised one brood, ere the

3

"crex, crex" comes from the grass field. It is one of the voices of the summer night, like that of the partridge. Although so generally distributed, and so familiar to the peasant that it seems one of the presences of the country, it passes some districts over, or only appears rarely, and in small numbers. Unlike the lapwing, and even the partridge, the instances of the corncrake wandering, and nesting beyond the fences must be exceptional. For one thing, the short herbage of the moors, and wastes would scarcely suit its hiding propensities.

Round this bird the war about migration hotly raged. Reluctant to take to the wing, it was hard to imagine that it would attempt a long sea voyage; and so it was supposed to creep into some sheltered corner, and sleep through the cold weather. Nor was the fact that it was never found, sufficient to disabuse the mind of this conveniently easy solution. Ignorance dies hard. But lighthouse records, and the habit of lighting on ships, in its passage, were incontrovertible.

When at length "cuck-oo" is heard mingling with the melancholy cry of the wood-pigeon, the wave has passed for the season.

EARLY BURN FISHING

HALF a mile of pleasant lane, beneath the budding trees, and between the budding hedges,
leads to my favourite burn. It is just such a path
as one is tempted to linger in; especially when the
birds are in full song, and every other one seems
to have a feather in its mouth. The beat of the
water-wheel is audible half the way.

There is an advantage in having the burn thus
near. It is so irritating, after a walk of eight or ten
miles, with the prospect of a similar distance back,
to find that the meal miller shuts down the sluice,
with the view of nursing the water in the dam,
just as the rise is coming on; and keeps you sitting
on the bank, among the buttercups and daisies, for
a good two hours, until he sees fit to lift it again.

I watch my opportunity after a little rain has
freshened, without unduly raising or dirtying the

water, or when sunshine and shadow alternate without either gaining the advantage, or should the breeze blow fresh enough to put a ripple on the pools. And, notwithstanding those favourable appearances, and the expectations they excite, should the miller prove awkward,—although friendly millers are more obliging than strangers,—I am still within easy reach of home.

I can walk down the lane in the evening, when the blackbird is at his vesper, and, fishing or no fishing, spend a pleasant hour on the cool banks, not only without exertion, but with considerable benefit to my night's rest; returning in the twilight when the blackbird is uttering his chirpy scream.

For years, all connection with the sea has been cut off by several miles of impure water; and that, perhaps, makes the fishing so much the more restful. All its suggestions are of green fields, and drinking cattle, and yellow iris, and the drowsy hum of innocent flour mills. Not that the life is without a certain variety. In addition to the pale-fleshed brown trout, the red-fleshed seem to find their way from the lake above.

The tint of the flesh is, probably, only a matter of feeding. A largely crustacean, or molluscan diet

seems to produce red-
ness. In creatures so
variable as trout, if species
were to be made out of
every little difference of
this sort, a book would not
be sufficient for the record. We might have light
trout, dark trout, and piebald trout; red, pink, and
white-fleshed trout. There are all these shades and

gradations, besides many more; and the same trout may be one thing one day, and another the next.

The safer plan is to fall back on the brown trout, and to group all the others, found in stream, or lake, round this, not excluding the Salmo ferox of the loch. Even the sea-trout crosses freely with the burn-trout, and, when shut up in fresh water, takes on the appearance of the other.

A few have left the tiny pools and purling currents, where the soft, unstimulating food was washed out of the bank, to spend awhile in the bigger watery world of the lake. There they have lain at ease in the still weedy depths round the edges, feeding, and fattening on the shell-fish clinging to the swordlike sedges. And now they have returned, so transfigured that the angler does not know them again.

The earlier of the stone flies are already beginning to crawl up the iris leaves, and the stems of the reed grasses; while the rest await their speedy apotheosis at the bottom, meantime providing an inexhaustible supply, in the form of caddis.

Not that the rush of insect life has begun; other forms lie down there yet imperfect; among them that grandest of stream insects the May-fly. The water will have to be warmed by a week or two's sunshine, to give them wings.

Let no one suppose that fly-fishing in a burn is boy's sport; if it be, then it is the sport of boys who are afterwards to become experts. It is a school in which all ought to begin. True, it admits of a light rod, and a short line, and that is a consideration to the immature. But the absence of straight runs, the erratic course, the sudden bends, the narrowness, and the bushes and grasses on either side demand a precision uncalled for in opener waters. The delicacy needed for such work once acquired, the strength for wielding a heavier rod, casting a longer line, and lighting the flies like gossamer on stiller water will come after. Besides, one learns the elementary lessons, that, however noisy the stream may be, he must not tread too heavily on the bank ; and, however broken the surface, he must never have the sun behind him.

It is the whole that charms. Scene, and circumstance find their way in, and, blending there, become a lifelong possession. The birds of the burn are characteristic.

The wagtails are back. They have never really been away; at least not all away. It is the grey species that haunts the stream-sides, and, seeing

that his tail is fully an inch longer than that of the
other, he is the "water wagtail." Species making
in wagtails seems to me to be as recklessly
indulged in as in trout.

In two great graceful curves, the bird spans the
distance from the sluice, where the water enters
the mill-lade; and lights on a miniature sandbank,
formed by an eddy when the stream is in spate.
For the thousandth time I have reason to admire
that ineffable grace of movement, that harmonious
blending of soft greys with bright yellows, which
together mark her out as the lady among birds.
She stands for a moment, still vibrating; and, then,
runs with mincing steps to the water. Scarcely
wetting her dainty feet, she picks up the insect
larvæ exposed in the shallows, or brought within
reach by the tiny wavelets, which advance, and
retreat, ever so little, in rhythmic motion.

With a straight flight, or one which only varies
with the bend of the channel, the water-ousel comes,
steadily, up the stream, and lights on a stone in the
very centre. There he bobs, like a great robin or
wren, and then, descending, he turns up his tail
and disappears under water, emerging, after an
interval, a few yards farther down.

There is no very apparent change in his coat

since the winter. He seems always to dress in the same sober suit of somewhat rusty black; with an expanse of white in front, not remotely resembling a dress shirt. If the grey wagtail is the lady of the stream, then he is the dapper little gentleman.

I have seen him in quieter southern waters, where he formed no conscious, or essential part of the picture as he does now; no artist would have put him in as a finishing touch.

The kingfisher seldom visits my burn. Its surface is too troubled, its bed too stony, its inhabitants too alert. To get a bare living on minnows, would tax his utmost skill; and trout are impossible.

The elvers, or young eels are in the shallows. The popular idea that horses' hairs cast on the water become vital, and wriggle about in this way is by no means extinct. This is not true; nevertheless, there is some mystery in the matter. The large eels are seen going down the stream in autumn, to spawn in the sea. In countless numbers, the young rush up in the spring. All that happens between is a blank. Assuming that those little creatures were born at sea, they must have come through miles of sheer poison, in which no trout would have lived a moment. The mortality must be considerable, since the angler is seldom troubled with eels. The wonder is that any survive.

The fish rise eagerly, and grip as they seldom do in later months. Give me a spring day, just after the willows, and birches have done flowering, and before the beech, and chestnut have shaken off their brown husks, for fishing. Soon, half a dozen are lifted on shore.

Nothing is prettier than a trout, before the sheen dies out of its skin. The spots are peculiar to the fresh-water species, or only shared by migratory forms before they pay their first visit to the sea. There must be some reason for this, as there is for everything. Even the conspicuous stripes on a zebra's back are said so to blend into one another in the twilight, when beasts of prey are about, as to make it invisible. The same holds good of the grey shades, and stripes of a wild cat. These spots may have something to do with the proportion of bright quartz pebbles in the gravelly bottom, or the points of light left by the trembling leaves overhead, or the waving grasses on the bank. In deeper waters, the spots scatter out into crosses, deceptively blending with the lights and shadows of the current, or the gentle riple caused by the wind.

" The salmon's back is fenced with tiny blue slates

like the miniature roof of a house. Could anything
match more exactly the blue slates with which
our rapid streams abound?" Were it not truer to
say, that the "new run" salmon wears the double
livery of the migrant, adapted to both spheres. The
glory soon departs, and he takes the muddier hues
of the fresh water. Sometimes he covers himself
with red, and black spots, like a gigantic trout.

The marine forms of the shadowless sea are
silvery, with a darker shade on the back, and generally without slates. If river forms took to the
salt water, they would put off their spots as of
no further use.

Examples of the brook-trout have been found,
on emigrating to the sea, as a rule to which there
are exceptions, to assume the brilliant silvery hues
of the migrating Salmonidæ, as well as the cross-
shaped black spots. Mr. Harvie Brown remarked,
June 12, 1852, on having caught at Durness
several so-called sea-trout, from a sea pool or first
pool at the mouth of the river, fresh water at low
tide, salt, or brackish water at high tide. From
their silvery appearance, they are known as sea-
trout; but are the river form, acclimatised to
brackish water, or else periodically visiting the
same between tides.

I have just unhooked the last, and am preparing to cast again, when a water-vole leaves the near bank, and begins to swim across the pool. This is the otter of the burn, but only in appearance; although the miniature is sufficiently deceptive. The otter would be out of proportion; but this little black creature, so perfectly at home, seems as if modelled for his smaller domain. He is maligned, but that is all prejudice. He swims among the rising trout, and they do not seem to mind. He dives, when he sees me; but not to follow the retreating fish, only to escape. I can see him under the water, until he reaches the opposite bank, and disappears below the trailing grasses.

He is not the black rat some people take him for: which creature is nearly extinct. He has the blunt nose of the vole, and is really as innocent as a guinea-pig. He is not assertive; he is retiring; competition would be the death of him. He has simply found a place of his own, which no other creature cares to dispute, and a very desirable kingdom it is. He thins out perceptibly by the mill there. The reason is obvious.

The brown, or Norway rat is in possession, and will admit no rival. He does not go up and down the stream, or scatter over its quiet stretches.

There is too little for him. He does not believe in vegetable diet, or hunting too hard for a living. He likes the neighbourhood of houses, and plenty; and nothing suits him better than a meal mill. How fat, and well fed the fellow looks!

What an unutterable rascal he is! It is he who has well-nigh exterminated the black native rat; whenever he enters a house he kills, or drives away every mouse; and, if he leaves the vole unmolested, it is only because he would not thank him for his quarters.

Perhaps it was the recollection of an early experience, that caused me to pause in my cast. One day, when fishing, as I am doing now, I allowed my line to drift under the bank, and hooked something living. From the bend in the rod, I judged that the trout must be immense, some patriarch of the stream, which, for the moment, had remitted his caution. The excitement changed its complexion, for I was younger then, when, half coerced by the strain, a water-vole emerged into view. After an exciting struggle of the pull-devil, pull-baker order, he broke loose. Once more I breathed freely, and so did he.

The typical Scots stream has really four chapters,

when its history is fully written out. It starts
away up on the wilds as a moor, or hill burn,
siping through the grass, forming little moist
patches, where grow the sphagnum and the bell
heather, and, finally, cutting for itself a channel
among the peat. The channel is narrow, the
stream shallow, except where it gathers into a
little black pool; but, every boy knows that it
swarms with trout. These trout average about
the length of one's little finger, are rather dark-
looking, to blend with their surroundings, and
not over palatable. Still they are trout; and
all count. And the simple delight of that day,
with one's new half-crown cane rod, and the
lunch, mother has so carefully packed in the
basket, and the half-weird fellowship of mountain hares, and the dozens that are laid out, and
counted, oh! so many times.

The hill burn enters a lake, which, strange as
it may seem, it once made, and, small as it now
is, it once filled up. It is really the same burn
which enters at the one end as a head-water, and
issues from the other as a tail-stream.

The second stage is the burn proper, such as I
am fishing now. By this time, it has, probably, got
beyond the limit of the peat. The bottom is

cleaner, and the trout are prettier; the feeding is richer, and the trout are less gaunt, and less harsh of taste. Many of them visit the open water, where they grow to a greater size, and may become red in the flesh.

The burn forms another lake, and issues as a " water." This name seems to be peculiarly Scots. The breadth is just as far as a skilful caster can send his tail-fly. Trout-fishing is at its maximum here. If the water is still, the fish become fat, rich, and lazy—too lazy often for sport, rising to several pounds in weight; if the water is rapid, they are long, muscular, and numerous, averaging a half, sometimes not more than a quarter of a pound. All things else being equal, the trout of still streams are the more delicate.

The water passes through another lake—three is the typical number—and issues as a river, where salmon reign, and trout take a second place.

Such is a brief account of most of the great rivers in Scotland—the Tay, with its tributary the Tummel; the Forth, with its tributary the Teith.

These four stages correspond with childhood, boyhood, youth, and maturity, and should prove an excellent angler's guide, Nature's hint as to the true course of development. He will be the most perfect in the art who has begun on the moor, and passed through the burn, and the "water" to the salmon stream ; and, he will be the best lover who looks back, perhaps with a sigh, from the salmon stream, to the early morning when he clambered up among the heather to the hill burn.

The course of my little stream is short. The chain, or watershed from which it flows is but ten miles from the coast, which its windings may increase into thirteen. It springs on the hillside, broadens into one lake, and enters the sea as a burn.

III

ON THE MOOR

A FRESHLY painted placard forbids the taking
of eggs, or even the going farther. This is
something new, and, as far as the eggs are concerned, in the right direction. A process of wholesale robbery had been going on, which threatened
ultimate depopulation. According to the late Mr.
Matthew Arnold, it is the blessed privilege of a
certain section of the British public to go where
they please, to do as they please, and to smash
what they please, and then to express no little
astonishment, and indignation when they are told
not to come back again.

It is even within the knowledge of some of us,
that genuine naturalists, who are loud in their con-
demnation when they hear of others taking rare
eggs, never lose an opportunity of securing them

for themselves. Were another great auk, whose extinction is so universally deplored, to be found sitting gravely on its one egg, there would be such an ugly rush as this planet had never witnessed before, and, it would go hard if the savants were not to the front.

Very mild and seductive the moor looks in the spring morning. A fresh breeze is blowing, and light shadows are pursuing the gleams of sunshine in a love chase.

Rabbits scud about in every direction. Those sandy-coloured children of the sand, instead of livening up the scene; by their silent movements and sudden disappearance, as if swallowed up by the ground, only make the solitude more appreciable. A single glance at the creatures, against the background, or environment, shows that they are protectively coloured. The assimilation is almost perfect. There is only one defect. The white tail becomes so conspicuous, when they are running away—and they generally are—as to challenge the attention of the most unobservant. Failing to understand why a harmony, otherwise so perfect, should be disturbed in this way, Mr. Darwin wrote to Dr. A. R. Wallace for an explanation. The latter suggested that, being mainly twilight feeders, the tail might be useful as a signal to the young, especially in time of danger.

The birds do not come quite so near the edge of the moor. I pass straight across seaward, and have gone almost half a mile, before meeting with anything less common than the lark, and the meadow-pipit.

At length, the familiar chat of the wheatears is heard. These too belong to the moors of Scotland, the downs of England, and the deserts of other lands, and, by their monotonous click, give voice to the featureless desolation. It would seem as if each living creature became a part, and in a degree an interpreter also, of the scene it inhabits.

One has to look, very carefully, in the direction of the sound, before he can detect the bird, so harmoniously do its colours blend with the background. But, as soon as it rises, the difficulty vanishes, and a strange thing appears. When it lifts the wings, it displays a white rump, which makes it even more conspicuous than the rabbit. The unexpected flash causes one to look; and it must be a very blind hawk indeed that fails to see.

It does seem a singular coincidence, that the mammal of the wilds, and the bird of the wilds, should have a similar mark. Will the same ex-

planation suffice for both? And, must we conclude
that the white rump of the wheatear is meant to
show the young the course the parent bird has
taken across the moor? It is not a twilight feeder,
but then, the protective colouring might hide it
even by day.

The wheatear reaches the Scots moors—for it is
a migrant, as early as March. The apology for a
nest, with its faint blue—almost white—eggs will
be in many of the disused holes of his comrade the
rabbit. They are easily found, because of his
slovenly habit of leaving chopped pieces of bracken
round the opening. As in the case of most of our
hardier migrants, a few may remain with us all
the year round.

"Cuckoo, cuckoo" sounds from a clump of trees
in the centre of the moor. The males arrive first,
and seem to affect the high ground, where they
call for awhile before scattering over the country.
Since the mimicry of eggs can scarcely be changed
at will, they are bound to seek the haunts of the
birds they have been in the habit of deceiving.
For the nests they search with the skill, and
persistence of a schoolboy. Should everything
else fail, they watch the bird going, or returning.

On the moor, the chief victim is the meadow-pipit. In arranging collections, I have found it hard to tell a cuckoo's egg from a meadow-pipit's, except from its greater size, so completely does nature enter into the conspiracy. The hunt for titlark's nests, therefore, and the scanning of the eggs, is always interesting, and exciting, when cuckoos are about.

The meadow-pipit is entirely a moorland bird, confining himself to untilled wastes, and the region of natural grasses. He comes midway between the lark of the cultivated fields, and the rock-pipit of our sea-coasts; sharing the wilder half of the former's domain, and just bordering on that of the latter. He is known as the moss-cheeper, and is credited with the lapwing's art, of making a great fuss at a distance from the nest, to mislead the searcher.

A shrill cry, very distantly resembling "curlew, curlew," varied by a loud and quavering whistle, rises above all other sounds. Generally, the whistle is used when at rest; "curlew" when on the wing. I am approaching the domain of moorland birds.

The species occupy certain well-defined zones. Entering from the landward side of a coast moor like this, one expects the first greeting from the lapwing, and the second from the curlew; crossing

the centre he is in the domain of the dunlin, and
the golden plover; approaching the sea he has to
endure the irritating attentions of the redshanks
and the terns; and if he goes a little farther, he
disturbs the ringed plover from her nest among
the gravel.

Other forms are scattered, but after some
definite plan, so that he who is acquainted with
their habits, knows where to look for them. In
the clumps of wood nests the owl; in the patches
of heather, the grouse, and eider-duck; in the
marshy places, the teal, and the mallard; in the
rabbit - holes, the sheldrake, and the wheatear;
behind the tufts of moor grass, the lark, and the
meadow-pipit; in the nest of the latter, or both,
the cuckoo.

There are, of course, exceptions to all rules, but
terns are not common on the landward side of the
moor, and curlews are seldom, or never found nesting on the wild half-naked sand-dunes, through
which the binding lime-grass is threading its
network of stolons. It were as vain to search for
the white eggs of the sheldrake among the heather,
as for the olive-coloured eggs of the eider-duck in
a rabbit-hole.

The first find is a lapwing's nest, with its four

darkly-stained eggs, blotched almost black at the
thick end.   As in the case of the other plovers, and
their allies, the eggs are large for the size of the
bird—that of the curlew is half as big again as a
hen's; that of the small golden plover
twice the size of a
wood - pigeon's.   In
shape, the eggs taper
off to a blunt
point, so as
to take

up the least possible amount of room.

These early builders are not wholly moorland
birds, but, like the lark, belong half to the areas of
cultivation.   Probably they began on the moors,
and overflowed into the fields.   Those reared on

the farm return to the same field, or, should the
crop be unsuitable, to those immediately adjoining;
those of the wilds revisit the wilds.   Without
implying any difference in structure, it might be
legitimate to draw a distinction between field, and
moorland lapwings.

The sitting bird rises, noiselessly, and flits away
like a shadow, while the male remains silent, and
invisible, both thus displaying greater wisdom than
the species get credit for.   An imaginary lapwing,
of the Sandford and Merton breed, would have
screamed and hung his wing, hence 'lapwing,' by
this transparent ruse attracting attention to the
nest, whose site it is his interest to conceal.   The
greater the danger, the more the fuss, and the noise
he should make.

Various species, from the mire snipe to the blackbird, hang the wing to draw attention to themselves; but wiser than they, the lapwing makes all
the fuss when the intruder is farthest from the
nest, relapsing into silence when he approaches near.

Six heavy birds flap and circle, with outstretched
necks, and level flight.   Three males are black and
white, and three females sober coloured, and all are
eiders.   There is no other duck with male and
female shaded thus, except perhaps the smaller

king-duck, which is much rarer. Judging from notices one sometimes sees, it does not seem to be generally known that this bird is a resident of Scotland, at least of such a low latitude, and builds on the seaside moors. "The eider-duck," says St. John, "is a rare visitant to this part of the coast, (Moray). It breeds in some of the more northern rocky islands of Scotland, though even there it is now rare." And here are whole six all at once, evidently, with no intention of seeking further for summer quarters.

For some minutes I have been walking through a stretch of heather, when an eider-duck rises almost at my feet, exposing a nest of four eggs, and thus settling, in the only practical way, the question of its breeding. According to her wont, she had trusted for protection to the unobtrusive colouring, which so exquisitely blends with the surrounding shades; and, only when she saw that I was coming straight upon her, did she rise, more quickly than I could have conceived possible in so heavy a bird.

In crossing the moor last season I came upon four sitting eider-ducks, evidently prepared to let me pass, however near I approached; and all of which I could easily have secured with the insect

net I had in my hand. In one case, I stood by the
nest for some moments without result, so assured
did the bird seem that safety depended, more on the
heather tinting of its feathers, than on flight.   In
all cases, where the female is sitting, the bright-
coloured male is out of sight.

The nest is interesting, and almost exceptional,
inasmuch as the dried grass, generally used as a
mere framework to hold together the soft bed of
cider-down, composes the greater part. So early
in the season the down is usually three-fourths of
the whole, burying the framework out of sight.
In case of a second or third nest, the proportion of
down diminishes.

The redshank has the lapwing's habit of flying
round, and piping at the intruder. The eggs are
very sharply pointed, and have a delightfully
cream-coloured ground with small brown spots.

A peculiar querulous half-threatening scream,
which once heard is never forgotten, coming down
from overhead, announces that the terns, or sea-
swallows from the south have reached their
breeding-ground. If any doubt remains about the
cry, there is no mistaking the jerky and yet buoyant
flight, the long crescent-shaped wings, the deeply-
forked tail.

So great are the numbers, so shrill the sound, and so keen the look down upon us, that one can scarcely wonder if nervous people have imagined they meditated an attack, and have even been known to seek safety in flight.

I sit down for a little to rest, and drink in general impressions. A patch of beauty, some two hundred yards away, resolves itself, under the glass, into a pair of sheldrakes, with their blue heads, and the exceeding purity of the prevailing white, relieved by streaks and dashes of black and brown. It is always easy to recognise this species when there is a pair, seeing that the female is equally bright with the male; a condition of things not common among the ducks, and probably connected with the habits of nesting. A bird that incubates in the shelter of a rabbit-hole can afford to be bright, and, for the same reason, her lord can safely, as in this case, take a share of the sitting.

Whereas the eggs of the eider-duck are four, occasionally five in number, and of a light and inconspicuous green, those of the sheldrake are from ten to a dozen, and white. The eider-duck's eggs are protectively shaded like the female. Sitting sheldrake, and eggs alike are concealed, and stand in no need of shading. The difference in numbers

may indicate in this, as in other cases, that the
sheldrake's eggs are exposed to greater risks.
The birds walk over the grass rather than waddle;
and altogether their movements are freer, and their
appearance more graceful than those of the other
ducks.

Flat as the palm of the hand, except where a
low ridge of crumbling sand-dunes shuts out the
view of the sea, the moor flows away on all sides.
It still wears the sober livery of winter. The
heather is dark, the grasses a faded yellow, the
sand, where exposed, tawny. A shadow on a moor,
as it darkens in succession heath, and sorrel, and
moor-grass, and shelly mound, and leafless trunk,
and dark fir-copse, and picturesque sand-dune, is
different from a shadow anywhere else. A darker
cloud, with just a few drops of rain, gives yet
stronger contrasts, and intenser effects. It is easy
to understand now why a painter should find the
scene at once so alluring, and so disappointing. It
eludes the utmost refinement of brush, or word.
One feels its power, without being able to reproduce
it; learns its secret, without knowing how to tell it
over again.

Not a flower, or fresh blade appears on all its
surface. A few scattered willows, a clump of trees,

whose dark Scots firs are hung here and there with
the tassel of the larch, and touched with the tender
green of the budding birch, are the only signs of a
spring, which, elsewhere, is already fast merging into
summer.

Here and there on the moor are placed certain
uncannie, gallows-like erections, which, in the moonlight, must give the furred, and feathered night-
wanderers quite a shock. These are meant at once
as perches, and traps for the long-eared owls.
What special crimes those interesting creatures
have committed, it were hard to say. They live
entirely on mice, shrews, and voles, and, from the
number of ejected pellets scattered about, they seem
to be extremely industrious, and successful hunters;
not that the mice do much harm on the waste,
but still they are a prolific race, like the rabbits,
and most other rodents, and want keeping down.

To know birds, one must not only be acquainted
with their feeding-grounds; still more important
is it that one should visit them thus in their
breeding haunts. They are gathered from mount,
and stream, and sea to the same moorland, where
they not only appear at their loveliest in their
bridal plumage, but, for the time, have laid
aside their native wildness. The fear of man is

swallowed up in a stronger emotion, and even the enmities among themselves are in abeyance. Carrion-crow, and kestrel nest on the same fir branch. The female sparrow - hawk blinks at the sitting chaffinch, and never thinks of picking her from the nest.

The forms visible on the moor include, in the order of size: meadow - pipit, lark, whinchat, and wheatear; dunlin-snipe, and redshank; merlin, and kestrel; golden plover, and green plover; cuckoo, lesser tern, Arctic tern, and common tern; curlew, and long-eared owl; teal, and mallard; sheldrake, and eider. All of these in considerable, some in very great numbers. At the bottom of the list comes the eider-duck, with perhaps a score of nests; at the top, the terns, with more than a thousand.

The grouse, lately introduced by the proprietor are spreading, and threaten a new incentive to the poacher.

IV

GULLS AND DIVERS

JUST over the sand-dunes, the terns are fishing. Their motions, while thus engaged, are well worth watching, and comparing with those of the osprey. We have only another ocean-diver from a height, and his diving is not of the same kind, nor performed so near the shore that every one may see.

Common, and Arctic terns are engaged in about equal numbers. They seem to mark out a certain distance, or beat, just beyond where the ripples of a quiet sea are breaking on the sand; and to pass over it again and again, presumably, as long as it will yield anything. They fish against the wind, even when there is so little as to-day; returning by a swift flight to the starting-point.

Their whole aspect is keen and intent to the last degree. Suddenly they pause, spread out their tails; and, with quickly pulsating wings, hover over the spot. Sometimes they rise a little in the air, as if to focus a blurred object, and then descend like a bolt. They have not the power to pursue their prey under the water, and trust solely to the force, and impetus of the dive. Their arrowy flight toward the moor tells when they are successful; which happens perhaps once in half a dozen dives.

The comparative certainty with which the terns can reckon on the motions of the sand-eel, which must be constantly shifting its position; and, though taking an appreciable time in the descent, dash sufficiently near the spot to seize the creature in its bill, is surprising. There may, of course, be a shoal beneath, which would broaden the mark, but that only lessens, and does not remove, the wonder.

The first unexpected impression, on glancing along the coast, and out to sea, is that of lifelessness. In imagination, we associate flashing wings with summer waters, and look for white plumage against the passing rain-cloud; and I am mistaken if some artists do not occasionally oblige us with that phenomenon. But, as a matter of fact, this is the most barren time of the year, and, but for those diving terns, would be more barren still.

All the bright life so
lately scattered abroad,
has been concentrated on the few breeding places — St. Abb's Head for the
southern group, Duncansby Head for
the northern group, and the isles of the
Forth, chiefly Bass and May, for the
central group.

With plenty of food all the year round, and no

5

semi-starvation season to live through, like their kindred on the land, the sea birds should increase indefinitely. Besides, they have no enemies, except an occasional peregrine falcon, or sea-eagle. A shark has been seen to rise and engulf a diver, and an ambitious cod to attempt a guillemot, just as a pike will sometimes swallow a water-hen; but the occurrence must be too unfrequent to affect the balance.

Where the risks are great, the precautions for the preservation, and continuance of the species are exceptional; and there is no better indication, and gauge than the contents of the nest. Twelve eggs, roughly, signify double the danger of six. Keeping this in mind as we turn to the sea birds, we find that the number of eggs is very much smaller than in the case of the land birds. This holds almost universally, with perhaps the exception of some of the ducks, whose nesting habits expose them to considerable danger.

A very common number is three. This holds throughout the gulls, and their immediate kindred. The terns' nests on the moor have three apiece. This is well-nigh the maximum. In the case of the skuas, it is reduced to two. The divers also lay two.

In the case of the petrels a limit is reached, beyond which further diminution is impossible. Indeed, the unit is so very frequent, that it may almost be regarded as the typical number. The guillemot, the razorbill, the little auk, the puffin, the shearwater, the solan-goose, all deposit and sit on one egg at a time; and they discharge the duty with more than the gravity of a sitting hen on her sixteen. The unit of production must be taken as representing the unit of danger, and means that the sea birds have so little to fear, that, in a majority of cases, one egg is found sufficient to maintain, or probably slightly increase the species.

Five is the usual number found in a land bird's nest. Four is not uncommon. Three is very rare, occurring with some owls, and hawks. Two is confined to the swift, and the night-jar, whose nests are very seldom found; and the pigeon, which builds out of reach of many enemies. And one is unknown.

Starting from five with the land bird, the tendency is upward, not downward; starting from three with the sea bird, the tendency is downward, not upward.

A few heavy-looking gulls, of a dirty brown plumage, evidently young, represent the vast armies of pure whites, and slaty blues that have departed. These are either last season's birds, or

may even be two seasons back.   Plainly they do not
come early to maturity—two, and even three years
being quite a common interval.   This helps to keep
down the rate of increase, and points to diminished
risk.   Did all the young nest the first season, the
number of birds must be vastly increased.   In
the case of land birds, the breeding is begun at once,
as if in feverish haste to make up for the immense
slaughter.   Were it not so, there is reason to fear
that many species would become extinct.

Familiar as they are to our eyes, and on our
tongues, the gulls are the least known of our sea
birds.   It is an excellent test of one's powers of
observation, and patience as well, if he can distinguish one from the other.   Of course, when the
initial confusion is overcome, a glance is sufficient
for identification.   The differences in their cry, in
their flight, in their appearance, become familiar.

The commonest of our coast forms is the black-
headed gull.   He seldom ventures inland beyond
the coast fields in winter; or far to sea, unless a
shoal of fish is driven in.   On a calm day he
takes a ride on the water as children on a swing,
sitting high, and looking rather awkward.   On a
rough day he searches the breakers, avoiding their
broken crests.   Usually he potters in and out,

watching, in that sharp way of his, to see that nothing is carried back again. He may be familiarly classified as our shore-gull.

His black, or rather brown head would be a sufficient mark, if it were constant. But then he puts it on in the spring, and off in the autumn; so that he is without it during the greater part of his stay at the seaside. In other words, it is his bridal dress. He does not really put off the old to put on the new. Therefore the rapidity of the transformation. The feathers simply change their colour. The effect is half a marvel. To-day he is pottering about in his white head, without any hint of what is to happen. To-morrow he appears in a brown, so dark that in the distance it looks black. It is as if nature had run a brush over him in the night.

Both sexes change alike. This is not uncommon among sea birds, and would seem to cast doubt on the theory that the male alone makes himself gay for the wedding. All it does show is that no explanation will take in every detail.

The change takes place about the twentieth of March. For a few days longer the freshly-painted gulls remain; but, by the end of the month, the sandbanks are deserted.

This gull nests in the reeds of some favourite inland lake, or marsh; beside the mallard, the teal, and the coot. These gulleries are often very large. Sometimes two or three pairs affect the rushes of some smaller sheet, where they have the company of the sedge-warbler, the whinchat, the white-throat, and the reed-bunting.

The peculiarity of the eggs is the diversity in marking. The ground colour is sober, and protective, the spots of the usual gull-like character.

The common gull is better known inland than by the sea, to the plough-man than to the fisherman. That may be the reason why observers on the coast report him as uncommon. He is really almost, if not quite, as numerous as the brown-headed species.

Any winter morning he may be seen leaving the sea-coast for his chosen feeding-ground—just about sunrise. Any evening, when the grey light of the winter day is swiftly passing into dark, he straggles back again from all directions. Anyone who came to the coast in the interval would naturally pronounce him uncommon.

This is why he is called the winter-mew, and one imagines that the name has been given by a landsman.

He makes no change in his coat, except to put off some dark spots, which he wears in the winter, from the crown of his head. He is a dumpy bird, with light greenish legs, and a habit of sitting all in a heap as if he had no neck. He breeds, by preference, on some flat seashore; and, occasionally, resorts to an inland loch.

The herring-gull is the largest of the grey species found on our sea-coasts. In addition to his greater height and freer gait, there is a wildness about his cry which makes it one of the voices of the sea, and a purity in his plumage which has gained him the name of silver-gull. Although he visits the fields, I prefer to think of him as he sits on the remoter banks, beyond the danger of surprise, or rides out on the bay after the shoal. He is the fisherman's gull.

If the herring-gull is the greatest of the common species, the kittiwake is the least. He is more of a sea bird than any of the others, neither visiting the land for food like the common gull, nor for breeding purposes like the black-headed gull. His zone is just beyond the breakers. In the water he is known by his smaller size and purer plumage; on the land by his clumsy motions and black legs; in the air by his buoyant flight and characteristic cry of "kitt-ee-a, kitt-ee-a." He is, perhaps, the most local of the gulls, absent from some portions of the coast, and abounding in others. He is, therefore —like many another creature—rare or numerous according to the position of the observer. I find a disposition among fishermen to call every gull a kittiwake.

In addition to the grey gulls, we have the two blackbacks. The lesser, though nowhere absent, is thinly scattered, and not nearly so common as on the west coast, where he breeds abundantly on the moorlands and lochs. He has an evil reputation for destroying eggs.

Not contented with the eggs, the greater blackback attacks the young birds, and is even charged with destroying lambs. Singly he sails into our estuaries, attracting the attention of the least observant by his great size, and the stretch of two yards from tip to tip of the fully-expanded wings.

The gulls grasp their food with the bill or feet indifferently, whichever is handiest, and nothing seems to come amiss. When fish are not available, and some do not put themselves to the trouble of seeking for them, they readily take any stranded or floating material. Although this omnivorous habit of theirs has its uses, one cannot help feeling sorry that birds, of which better things might have been expected, should act as scavengers.

The blackbacks are noted offenders. It is quite common to see " the lesser " feeding along with the carrion-crow ; and even "the greater" will stoop to a stranded, and not over savoury fish.

Really, they are not so very much to blame, seeing that they cannot very well help themselves. Not being able to dive, they are forced to take whatever comes within their reach. The surface of the sea is proverbially barren. And as its inhabitants do not as a rule, come ashore of their own free will, and alive, but only when they are washed up, the gulls are forced to take them as they find them.

For such fresh fish diet as comes their way, the gulls are dependent on any fry that may be scared to the surface. There are times when the barren sea is fertile and springing with life. When the herring sile comes inshore, nothing can be more interesting than the attendance of sea birds.

The scene is animated in the extreme, and illustrates, in a dramatic way, the functions and

limitations of gulls and divers.   The last I saw was
enacted in a patch of sunlight, shot from behind a
cloud on a grey sea, and was as bright as it was
lively.   The prevail-
ing   white   of
the breasts

w a s
almost
dazzling in its purity.

Great solan-geese rose high in the air, and shot
down again into the water.   Under the glass, the
divers, including guillemots and razorbills, could
be made out disappearing and rising in the busiest

fashion; while a mass of gulls, in a state of the utmost excitement, seemed to hide away the sea.

The divers are at the opposite extreme. When the gulls leave the surface, they rise into the air; when the divers leave the surface, it is to go down into the depths. There are intermediate forms which fly or dive indifferently, but the divers proper seldom rise if they can help it, not even when they are hard pressed. Wonderful is the adaptation of structure to life. The upper half of the leg, which might be of use in walking, is hidden away inside. When they come ashore, which they only do to breed, they shuffle along their body, which rests on the ground, in the most awkward manner imaginable. The exposed half of the leg, which is carried far back, is almost as sharp as a knifeblade, and is turned broadside on for the back stroke.

There are three such divers on our coast. The Great Northern, as his name implies, is the most nearly Arctic form, breeding sparingly indeed in the north of Scotland, but not in any great number till we reach Iceland. This is the "loon" of the whalers. After nesting, he returns south, and is by no means an uncommon form around our coast. He can hunt for almost a quarter of a

mile under water without coming to the surface;
and his speed there equals the speed of many birds
on the wing.   He is occasionally entrapped in the
salmon-nets, or takes the bait on the fishermen's
lines.

The blackthroat is the finest, and also rarest of
the three.   He is known to build on the grassy
islands of several Scottish lochs, especially in that
paradise of wild-fowl, Sutherlandshire; but he has
been too much run after to increase.   Probably he
is on the decrease on the east coast, being now
seldom seen.   He is said never to take wing.

The most southernly, and commonest is the redthroat.   All three follow the shoals, and this one
has earned the name of "sprat-loon."   He is also
known as the "rain-goose," from a peculiar cry,
supposed to portend bad weather.   .

A fourth has been recently added, called the
whitethroat, and even the "Canadian diver"; and
any winter, or spring day he can be seen anywhere
on the coast diving just beyond, often within the
broken water.   But observers failed to note that
when the whitethroat was there, the redthroat was
absent.   Nor did they reason that, like similar
strange marks, the red probably belonged to the
breeding season, and when absent would neces-

sarily leave the throat white. It is just as if the black - headed gull were called white - headed in winter, and two species made out of one.

It is quite possible that similar mistakes may have crept into our natural histories. The weakness of all inquirers is to discern something new; and, in their eagerness, not to take sufficient pains to verify their conclusions. Before we make additions, it might be as well to revise those we have already made, to assure ourselves that we are not mistaking phases of development from the nestling to the adult, or seasonal changes of plumage to meet the exigencies of the weather, or finery put on for the wedding, as quite another creature. This would still be discovery, entitling the investigator to have his name in the papers, with the further decided advantage of being in the right direction.

To solve the mystery, a friend went out to shoot a whitethroat, and found the task he had undertaken of no ordinary difficulty. Unlike the buoyant gull, whose whole body seems to be out of the water, and which could not get under the surface if its life depended on it, the diver is so low set that the wavelets lap over his back, until only the neck is visible. Between the firing of the shot, and the scattering of the pellets on the water, the bird had disappeared, to rise again unhurt, and out of range. When, at length, a long chase ended in success, the red was seen to be already blushing through the white.

V

THE NORTH SEA

THE breath of the North Sea, rude, but kindly, meets me, blowing coolness and energy through every channel of body and mind. "I have never known what freshness was since I left," said one who spent a month here, and was hastening back again.

There is no resort, not the shadow of woods, or the loneliness of mountain-tops, more delightful to me than the belt between the tide-marks, flushed twice a day by racing waves. Cool olive blades of the larger seaweeds float on the surface of rocky basins, richly inlaid with the pink of coming coralline; while bright and dark greens, and reds, form miniature forests in the shallower pools.

Every here and there, anemones peep out from amid the abundant leafage. Fancy, running riot among the varieties, as if in search of sufficiently

79

beautiful names, calls them rose, and dahlia; daisy, and marigold.

The common shore species are the dahlia (*Tealia crassicornis*) and the less attractive beadlet (*Actinia messembryanthemum*). But, out in the bay are the nobler dianthus, flower of the gods; and a still larger form, usually found journeying about on the crab's back, till it overlaps the shell, and makes the life of its host not worth living.

I edge up some of the stones, careful lest the drip dim the surface, and the eel-like gunnel, or the viviparous blenny, the bearded rockling, or the formidable bullhead, the quaint Montague-sucker, or some transparent gobby, darts, or wriggles away.

Abandoning the vaguer waters outside to his larger relatives, the lesser tern is poising over the pools, focussing some living form which has ventured from the shelter of the weeds. I watch, in fear, lest he strike the side of the rocky basin, or dash himself against some stone in the shallows. But still he rapidly drops, and rises again, and flits on to the next. No one, who is unacquainted with the bird, can conceive how exquisitely delicate, graceful, and even spiritual he is. I have seen many a perfect picture—nature is full of them—

but never one to equal that fairy bird, quivering over these fairy oceans.

I join the boys who are fishing out at low water-mark; dangling their bare legs over the furthest exposed rock, paddling their brown feet in the water, and dipping their not very delicate bait down among the thousand discs of the great seaweeds, to tempt the not very particular red codling, and dark poddles.

This outer limit of ocean vegetation, known as the laminarian region, seems to be used as a sort of nursery for the young of many of the larger fishes, which, erelong, will go out to the great watery world, and engage in the fierce competition constantly raging there.

This domain has long been known to lovers of nature, and also to those who, after a certain dilettante fashion, were also students. Many books have been embellished, and many aquariums stocked, and much poetry written, much delight experienced, and a little knowledge gained.

It is only recently that the sea, beyond the lowest ebb, has been systematically scanned and explored; and the variety, and ways of its inhabitants made as familiar as those of the rock pool. So long as the investigations were left to

6

private enterprise, however enthusiastic and intelligent, the information could only be fragmentary, and the results could scarcely be expected to command wide acceptance. But a new development has taken place, and the work has fallen to more competent, if not more disinterested persons. The aquarium has increased into the tank, the naturalist has crystallized into the scientist, and delight we are afraid has solidified into duty.

The centre of this new activity is the marine laboratory, quite a modern institution, whose existence is not generally known, and whose value is not yet sufficiently recognised. The instruments of research are dredge, and net. It has been my privilege, during several seasons, to share in the work of such a place, and to watch the yield of the sea.

Most of the marine invertebrates spawn early in the spring. Although themselves fixed to the rock, or crawling along the bottom, or burrowing in the sand, the young forms—young starfishes, urchins, annelids, ascidians—enjoy for a time a free ocean life. Very beautiful all of them are, especially those which move by rows of cilia, revolving like minute wheels. These larvæ, though by no means the exclusive, are the characteristic forms secured in the net between February and May.

Then come the medusæ of our summer sea. These have got their name from the habit some of them have of twisting up their tentacles, like the locks of the fabled Gorgon's head. But the unpleasant suggestiveness is removed by some epithet such as delicate medusæ, or medusa flowers; and thus we are set at liberty to admire the beauty of the creature. Medusa flowers is perhaps the happiest description.

The ugly name jelly-fishes is only applicable to the larger medusæ stranded in unsightly masses, along the coast. That with the four purple horse-shoe marks is *Aurelia aurita.* The large brown mass with long streamers is *Cyanea capillata.* The latter is a formidable creature, because of its stinging capsules.

All these propel themselves by the opening, and shutting of the bell, or disc. Others move by the surface play of eight rows of flaps, whose direction is marked by the flow of waves of iridescent light.

The sanguine hues of life are toned down to the neutral shades of the surroundings. In the crystal waters, we have crystal organisms. The eye can look through and through them, and, when at length, not without difficulty, it separates them out as distinct objects, it can see the vital processes,

which are hidden away beneath the opaque exterior
of land forms, in all their marvellous detail.   It is
this that makes them so invaluable to the zoologist.
The processes are simple.   It was in the sea that
life began.   It is in the sea that we make our
nearest approach to the mysterious genesis.

In earlier, more unsophisticated days, these
things would have been regarded as interesting in
themselves.   But an utilitarian spirit is entering
into the age, in accordance with which, attention is
being increasingly directed to the economic, and
practical aspects of marine zoology.   To the sportsman, as we shall see hereafter, all land animals fall
quite naturally into the two divisions, game and
vermin; and, in like manner, from this new point
of view, marine life divides itself, quite as naturally,
into food fishes and their food.   .

The arrangement is delightfully simple, and
should save one a world of trouble, consisting as
it does of the two easily remembered gradations:
what man eats, and what man's food eats.

These forms, which, at one time, would have
been known as flowers, or gems, or by some other
pleasant title—happily the names will stick—have
been slumped together, and re-christened the larder
of the sea.

The higher types, which satisfy modern conditions of importance, and fatten on this teeming population, may be divided into round, and flat fishes.

Roughly speaking, proceeding from the shore, the first we come to are the flat fishes; nor have we far to go. Tickling the wading children's feet are innumerable small plaice, from the size of a sixpence to that of half a crown. Farther out these remain numerous, but increase in size, and become worth fishing for. When mature, they go off to spawn, and do not seem to return. The limit is probably within a hundred fathoms.

Still roughly speaking, the zone of the round fishes is beyond that of the flat fishes; although, of course, the two overlap, and at certain seasons the former come close to the coast.

The round food fishes, with two important exceptions, are allied to the cod. The mere mention of the species into which it breaks up, is sufficient to indicate the importance of the family. The cod, the whiting, the haddock, and the ling, are among the more familiar, and valuable.

This is a northern type. Species of the family form one of the principal articles of food to those inhabiting the coast of the Arctic Ocean. Cod, haddock, and whiting grow to a great size off the coast of Iceland, whither our trawlers go in search of them. They abound round Orkney, and Shetland. They still retain their numbers, while diminishing in size and weight, in the chiller waters, and probably scantier larder of the North Sea. But, beyond the south of England, they thin out more and more. Their range is from 75° within the Arctic Circle, to 50° in the English Channel; and they seem to approach no nearer the equator than 40°.

With the flat fishes the order is reversed. Their headquarters are no longer near the Arctic Circle, but in the Mediterranean. They enter the North Sea through the Channel. The sole is perhaps the most southernly type, and the first to fail. It abounds off the coast of Devonshire. It is still caught in small numbers by line. as far north as Northumberland; but falls off rapidly after passing the Scottish Border. It is represented by a few scattered individuals along the east coast; and recent experiments, with a view to increasing the number, have as yet led to no result. It never reaches Iceland.

The turbot somewhat overlaps these limits. So does the brill.

Plaice and common dab (*Pleuronectes limanda*)

abound where the bottom is sandy, and may be regarded as the dominant forms of the North Sea, especially the northern part of it.

Lemon-dab are trawled in sufficient numbers on stony reaches. The flounder affects brackish waters, but comes to the sea to spawn, when it is found off the mouths of rivers.

The holibut is at once the largest, and most northernly of the flat fishes. Off Iceland it attains monstrous proportions, and is so heavy that very few are needed to make a ton. It thins out before reaching Aberdeen, very few are captured in the latitude of the Tay, and none at all past the south of England.

Details might be multiplied to confusion and weariness; but this seems to be the main feature in the natural history of the North Sea. It is a common water of the southernly tending round fishes, and the northernly tending flat fishes, where haddock, and cod meet with turbot, and sole, and mingle with dab, and plaice. For, although the flat fishes reach so far north, and there, probably because of an abundance of food, grow so large, it is still true that in high latitudes, where round fishes abound, they are practically unknown.

Probably, the English waters, being farthest south,

have the advantage in the flat fish; and the Scots
waters, being farthest north, in the round fish.

Among the semi-transparent larvæ, and other
products of the net, there appear, between February
and July, certain tiny globules, which so exquisitely
blend with the medium, that, but for a delicate rim,
they must be invisible even to the most practised
eye.

That these were the eggs of the food fishes was
first of all placed beyond reasonable doubt by
hatching them out in the laboratory. And, now
slight differences in size, the presence or absence
of oil globules, and certain other marks, are
sufficient to determine to what species each must
be assigned; even before the tiny embryo, with its
dots and stars of black, or brown, or orange, or red,
or yellow, wraps itself round the enclosed yolk.

By noting the dates of the appearance, or disappearance of the various species, and their relative
abundance or scarcity, and by carrying on the
process from year to year, the spawning seasons
are known, and the increase or decrease are
determined with approximate accuracy. The plaice
is our earliest spawner, and is followed by the dab.
"A January haddock, and a February hen," probably
indicates that the former is still unspent and in

good condition in January, and spawns shortly afterward. The actual time is the middle of February. The cod follows immediately, say, about the beginning or middle of March.

The most far-reaching conclusion is that the eggs of nearly all the more important round fishes, and also the food flat fishes, are freely suspended in the water, not very far from the surface, where they float separately, and at a widening distance from one another. The myriad eggs of a spawning cod, for instance, would rise and scatter, and each egg would take a course of its own. They are therefore not exposed to any wholesale catastrophe; and are safe, except from such unavoidable risks as tide, current, and marine enemies, which are by no means inventions of yesterday.

The actual risks to these floating eggs are quite sufficient without inventing new ones. Long before trawlers were heard of, and while the demand on the fish market was as yet of the modestest dimensions, nature met a well-nigh infinite loss by a well-nigh infinite supply. Before the first line was cast into the sea, a single cod produced its nine or ten million eggs, ripening them not all at once, but over an interval of six weeks, that they might not be exposed at the same time.

Whether there is any reason to fear that the margin is insufficient; and to add to the ten millions multiplied into all the cod that swim the sea, a few millions more, is very doubtful. Possibly any alarm in the matter is premature, and the anxiety wears a certain external resemblance to the act of Mrs. Partington and her broom. But, with a praiseworthy foresight, precautions are being taken.

As a development of the marine laboratory, an experimental breeding station has been started at Dunbar, where the eggs of the fishes—chiefly the plaice as yet—are passed through the earlier critical stages, and finally dropped into the sea. With what result to the yield of St. Andrews Bay, and the Firth of Forth, which have been the chief bene-fiters, it will probably remain to the end hard to determine. Upwards of thirty-eight million plaice eggs, and two million cod, have been distributed this year.

The main exception to all this is the herring. These gather to the banks from the north, south, east, and west, at the early and late spawning seasons. In summer they form such immense shoals, and are swept into the nets in such amazing quantities, as to make them the food of the poor.

The milt, and roe show the condition under which the capture takes place. When the object is accomplished, the assembly dissolves, or disappears as mysteriously as it grew.

For a long time these creatures were the mystery of the deep. Whence came they, and whither did they go? No doubt the migration of fishes is still less understood than that of birds, and a good deal of light remains to be thrown upon it. But there seems reason to believe that herring do not disappear as a body; but simply scatter, and pass the remainder of the year singly or in small shoals. The eggs of the herring are demersal, or sunken; and there are certain places favourable for their reception, whither they are led, as by an overmastering impulse, at the same time. Hence the shoals, which as quickly vanish.

It was long a puzzle how a creature so closely allied to the herring as the sprat should have floating eggs like the cod, and not sunken eggs like the herring; and, for a while, refuge was taken in denial that the eggs in question really belonged to the sprat. The easiest way out of such difficulties is to confess our ignorance, and inability to explain everything.

Indignation has been freely expressed over the

wholesale destruction of myriads of eggs—with all their possibilities of profit to the line fishermen, and cheap food to the people—by the passing of any heavy object over the spawning beds. And, if the spawn had actually lain where the objectors supposed it did, the protests would have been more than justified. But the evil, like many another, was imaginary. It is obvious that the trawlers can do little harm in this direction.

The Duke of Argyll said in the House of Lords, that, when one of the most eminent marine zoologists in Scotland promulgated the fact that the food fishes deposit their spawn in the open sea, and not at the bottom, the local fishermen were so angry that they burned his effigy. That was the way in which uneducated men, full of prejudice, regarded scientific investigation.

This mode of dealing with unwelcome truth is not confined to fishermen. There are those of more pretensions, who, not satisfied with the effigy, have disposed of the man. Let us be thankful that the naturalist in question is still amongst us. Give them time, and the fishermen, like their betters, will bow to the inevitable.

Probably the Duke did not state the case quite fully. Self - interest had more to do with the

innocent burning operation, than ignorance. Nor
are fishermen alone in crying out when their
livelihood is attacked, and the bread of their wives

and children
placed in jeopardy. But
the car will not stop
because it crushes a
few.

The Nemesis may come in
another form. So long as the fishing will pay,
trawlers will increase, and jostle each other on
the banks. There must be a limit somewhere, and

signs are not awanting that the limit has been well-nigh reached.

Picturesque ways of fishing must give place to other, and more efficient ones; just as the picturesque shearer gave place to the reaping machine, and the picturesque stage-coach to the railway train. No doubt the villagers, whom the six weeks' harvest enabled to lay in a little money against the winter; and the Jehus who were deposed from the box, called out in their turn. Grant that so many sons, and grandsons of fishermen may continue in the occupation of their fathers as to protect the three-mile margin, if only for the sake of the amenity of our coast. But that seems almost past praying for.

Perhaps, the world will lose as much as it gains. Nothing will ever charm us like the fishing-boats, with their brown sails swelling in the breeze, or rocking idly off the harbour in the morning, or evening twilight, waiting for the flow of the tide. The fish will never look so silvery, or seem so fresh again. But we can only sigh; we may not oppose. We live in an utilitarian age. Henceforward man shall live by bread alone.

MARINE MAMMALS, AND PREDATORY FISHES

NEARLY all fishes are predatory, in the sense of attacking, and consuming everything less powerful than themselves. It would be difficult to tell how many species follow on the trail of the herring. The larger members of the cod family turn from the hungry labours of the spawning bed, to get into condition again on herring diet. The cod itself has been fitly named the sea-wolf.

More eager still are the coal-fish, and the green-cod, known respectively in Scotland as the saithe and the lythe. These are the veritable Dromios of the sea, constantly confounded, even by the observant. If we may judge from the confusion in the minds of men, the fish themselves must sometimes be surprised when they meet: "Methinks you are my glass, and not my brother."

The former has a darker shade along the back, from which he gets the name of Black Jack, and the trace of a barbel under the chin; the latter is known by his protruding lower jaw. The coal-fish is much the commoner form in the North Sea, the lythe perhaps has the advantage on the west coast.

Whereas the professional freebooters of the deep are erratic in their movements; depart and return according to no rule, not even the abundance of prey; and may be absent for a long time without any apparent cause; the coal-fish is always on the scene, and ready for action; and, whereas many of the sharks, and their kindred, notwithstanding their bad character, are contented with lowly diet, the coal-fish exacts his daily tribute of highly organised victims.

Nevertheless, there is something in the equipment, and very appearance of the true predatory fishes, which unmistakeably marks them off for their vocation. The more or less pointed head, passing into the cylindrical body, furnished with ample fins, and ending in a powerful tail, give the needful speed in pursuit, while the wide gullet enables them to swallow large prey.

More characteristic still is the formidable array of teeth, fitted not only to grasp, but also to retain slippery, and fleeting prey. The fresh-water pike, whose more sluggish habit of rushing out from his den, under the alder roots, on passing trout, renders speed less essential, has still a similar arrangement of teeth.

The shape, together with the structure of the skeleton, admit of more than one bend in the body at a time. Thus the creature passes through the water with a rippling motion; which, albeit graceful as that of the snake, has something of the same sinister meaning.

The marine forms belong to the sharks, and dog-fishes. The transverse, or crescentic mouth under the prominent snout, seems to mark them out as bottom feeders, and necessitates the turning over in an awkward manner, when any object on the surface has to be grasped. It is quite possible that the disagreeable propensities, which make some of the species the terror of the tropical seas, and have come to be associated in the popular mind with the whole, was an afterthought, a habit painfully acquired, and still clumsily executed.

An interesting feature, not common among fishes, is that, in a majority of cases, the young are brought forth alive. The exceptions are the true dog-fishes, which are small round sharks; and the skates or

7

rays, which may be described as flattened-out sharks.
These enclose the fertilised eggs in the parchment-
like or sea-weedy cases, picked up empty by children
on the beach, and popularly known as sailors', or
mermaids' purses. The curious long horns at the
corners are used to bind them to some submarine
object.

In their passage down the oviduct, the ova receive
the case, much as the hen's egg receives its shell.
The female makes the first two horns fast by
moving round the mooring object, and then releases
the other two, and secures them at her leisure. A
slit on each side admits the free passage of water
through the case, and the inmates ultimately escape
by an opening at the square end. From the time
of deposition till the young pass into the sea, is
from seven to nine months.

The smaller spotted dog-fish, or the dog-fish with
the smaller spots (*Scyllium canicula*), is not at all
common along the east coast. The larger spotted
dog-fish, a deeper water form, still less so. Both
increase towards Orkney and Shetland, and continue round the west coast to the English Channel.
The North Sea, as will appear with regard to
other forms, is singularly free from pests of all
sorts.

Of the rays, the thornback (*Raja clavata*) is the dominant form, and the skate (*Raja batis*) is common. Their very shape indicates their proper place among the flat fishes at the sea bottom, and the clumsiness of their movements condemns them to such molluscan, and other diet, as they can find there. Should any fish approach near enough, it is speedily engulfed. The mouth of the rays being entirely on the under-surface, their prey is not directly seized with the jaws, but "the fish darts over its victim, so as to cover and hold it down with its body, when it is conveyed by some rapid motion to the mouth."

Among the sharks, the thresher produces purses like the dog-fish. The still larger Greenland sharks seem to differ from all the rest, in that they deposit the naked eggs in the mud.

The piked dog-fish (*Acanthias vulgaris*) owes his common name to the pikes or spikes, standing up like detached rays, in front of the dorsal fins. He is a viviparous form; and in this, as in other particulars, differs from the true dog-fishes. His voracity is unbounded, and the mischief he is capable of doing in a short time, incalculable. He comes down on the entrapped herring or pilchard—herring in the north, where pilchard is a rare form—

like a wolf on the fold, swallowing them up, and the net along with them.

Where they are numerous, the fishermen are often at their wits' end as to how to deal with them, and have been known to abandon their enterprise for the season. In Orkney, when I was there, four years ago, the seashore was simply littered with dog-fishes. A spirit prevailed that the wisest thing to do, might be to let them have the sea to themselves for awhile.

As some sort of compensation, no part of them is allowed to be wasted. They are slit up, and dried, and the rough skin used for cleansing purposes. The roof of the crofters' kitchens is strung all over with their bodies, for future consumption. No further preparation seems to be necessary, as the quantity of oil is a sufficient preservative. The liver yields combustible material to burn in their cruisies, during the long winter nights.

Twenty-five, or thirty years ago, the North Sea was infested with these pests. Old fishermen relate, that they were glad to see their lines safe aboard, and, in lieu of haddocks, were forced to be contented to barter a full cargo of dog-fish, for what they would bring. They crowded into the harbours, and could be seen skulking about, in

search of anything edible. Boys angled for them
from the end of the pier.

Then they disappeared, and have not. since
returned.   A few venture down the length of
Aberdeen ; scarce any beyond. The relief was great.
Perhaps no more signal change in the balance
of North Sea life has happened within living
memory.   The presence of the trawlers is as
nothing, compared with another advent.

Allied to this little dog-fish is the Greenland
shark, which grows to fifteen or twenty feet in
length.   Big as he is, even he shows the respect for
human life common to all our visitors.   Unlike
most sharks, which hail from the south, this is a
truly Arctic form, sending only a few out-swimmers
down our way.   He is probably the most sluggish
of the sharks, his fins being comparatively small,
and unsuited for swift locomotion.   When the
whaler abandons the stripped carcase, these sharks
gather in multitudes to the feast.

The so-called " hounds " have about the same
relation to the " dogs " as the greyhound to the
terrier, and owe their name probably to the habit
of hunting in packs.   They are comparatively
small, harmless, shore-haunting fish, never absent
from our summer seas. Two species are common.

The smooth hound (*Mustelus vulgaris*) is about four or five feet in length, and very mild for a shark. He lives largely on molluscs; or anything else of an innocent, and inexpensive nature that comes in his way. He appears on the coast in May, or June, swimming quite close inshore in search of food; so close, indeed, that he is frequently trapped in the salmon-nets.

The white hound, or tope (*Galeus vulgaris*), a much more spirited creature for his size, would seem to swim lower than the smooth hound, since he manages to escape the drift nets. He has been very troublesome to the fisherman of late, by taking the bait, and probably the fish as well. When hooked, he allows himself to be drawn near enough to the surface. But as soon as he comes within sight of the boat, and realises the state of matters, he snaps the hair, and disappears. No wonder that so many hooks are found inside when he is opened.

The blue shark (*Carcharias glaucus*), under favourable conditions, attains to nearly five feet, but is seldom found off our coast full grown, or of sufficient size to be dangerous to man. He confines himself to herring, mackerel, and other shoals; and would probably do so under any circumstances.

In warm weather he rises to the surface, and has the southern shark's habit of swimming with part of the dorsal fin and tail above water. A timid person, seeing this, would be extremely likely to get on to dry land as speedily as possible.

Two members of a family of large pelagic sharks find their way into the North Sea, with greater, or less frequency. One of these is known as the porbeagle (*Lamna cornubica*), which seems to be a corruption, or telescoping together of the words porpoise, and beagle, or hunting-dog. He is bulky, and thick set among the sharks, provided with large sharp teeth, and quite capable of doing mischief. Although he might not take the initiative in an attack on man, he has the character of being ready to act on the defensive. An instance is recorded by Day, of one which, when it found that it was captured, made a spring at the fisherman, and succeeded in making an ugly rent in his clothes.

The North Sea boasts another, perhaps more interesting, certainly more peculiar member, of the same family of large pelagic sharks. A thresher (*Alopecias vulpes*) was ignominiously shovelled into the trawl, off the Forth; doubtless as he was feeding on some fish at the bottom. Although sharks will take the hook, or occasionally plunge into a net, they are so difficult to get on board, that a capture has been an event causing no little commotion. The trawl promises to be a much more efficient instrument, not unlikely to increase the available supply, and reveal the presence of larger numbers in our summer waters than we were before aware of. Already they have lost much of their strangeness, and are becoming so easily procurable, that there is no reason why they should not shortly appear in our fishmongers' windows, as a substantial addition to our food-supply. An experiment was made here the other day, and the flesh was found to be as palatable as that of the flat shark, or skate.

The long claspers immediately proclaimed the sex; and, as among sharks the male is usually smaller than the female, his size of fourteen feet was considerable. The lower surface was a beautiful marbled white. The pectoral fins were large, doubtless to make up, in some measure, for the modification of the extremity into a weapon of offence. The upper lobe of the tail, which is about as long as the body, in this case about seven feet, tapers away in a whip-like fashion, with a thickening or projection at the end, which may serve the purpose of a knot in a lash. It is this lobe which he uses, with such irritating effect, on his huge victim.

Gunther says that his reported attacks on the whales is due to imperfect observation. To which Day responds: "As it has been abundantly proved that threshers do spring out of the water to strike down prey, I think it open to doubt whether all such observations are to be summarily dismissed, before the accounts are refuted, or another agent convicted."

Lord Campbell mentions a case of a shark leaping at a whale. Another observer witnessed several sharks similarly engaged, and when one was caught, and brought on board, it was gorged with blubber. Indeed, the reports of seafaring men leave very little room for doubt.

The thresher makes a still astuter use of his tail. He swims round the shoal in narrowing circles, lashing the water at intervals to drive his alarmed victims in a huddled mass to the centre, where they are at his mercy.

The sharks produce comparatively few eggs, or young. Against the nine millions of the cod, we hear of from thirty-two to fifty-two in the case of the tope ; of eleven young in that of the smooth hound, and of two embryos in that of the porbeagle. And, although eggs and young are matured from time to time, in no case can the number be otherwise than trifling. The hard cases of the eggs, and the fact of the young being born alive, doubtless help to protect them from many risks.

Side by side with those of the cod, the eggs are monstrous. So large are they, and so much food yolk do they contain, that the thrifty Swedes are said to find in those of the dog-fish, a convenient substitute for hens' eggs. I am not aware that they are so used in this country, but the hint is worth considering.

The step from the fishes to the marine mammals is a considerable one. It takes us from animals which breathe through gills in the water, and are cold-blooded, to warm-blooded and air-breathing animals. Moreover, the mammals, as their very name implies, not only bring forth the young alive, but also nurse them through the earlier, and more tender stages.

There seems reason to believe that these predatory mammals were once land animals, which, at some distant date, after a long period of terrestrial progression, and for some reason, probably connected with food-supply, took to the water.

Unlike the fishes, whose temperature more nearly approximates to that of the medium in which they live, they are forced to clothe themselves in some way against the chill. Therefore the inner layers of fat, or blubber, and the outer covering of fur; which have become such incentives to adventure, and valuable articles of commerce.

We, on the north-east coast, are so accustomed to all things northern, that we have almost begun to think of Greenland as an extension of our own islands, a somewhat more distant Orkney, or Shetland. The growl of the polar bear, and the yelp of the Arctic fox may be heard in the neighbourhood of our harbours, at the back end; and the Esquimaux has become a familiar figure in our winter streets. Our museums are stocked with every form of circum-polar life.

The smell of whale oil, hanging heavily on the breeze, is one of the disagreeable features of our acquaintance with the Arctic regions. But, albeit so familiar with the blubber and bone, very few of us have seen leviathan in the flesh. The cases are so rare and exceptional in which the right whale (*Balæna mysticetus*) has been known to come south, that he may almost be excluded from our ocean fauna. This attachment to the margin of

the ice-fields is undoubtedly a matter of choice;
for these Arctic waters literally flow with medusæ,
winged molluscs, and other items of his favourite
diet. The short space which he marks out, and
over which he swims backward and forward open-
mouthed, probably yields more than many a mile
of the North Sea.

The finner, a much less persecuted creature,
because of the shortness and comparative valueless-
ness of the bone, occasionally appears in the offing,
or even enters the mouths of rivers, and is the
species known to us as " the whale."

" In the cetacea, the dental germs are developed
in fœtal life, but the teeth either fall out before
birth (whalebone whales), or develop into permanent teeth (dolphins)."

The quiet life of our northern parts is sometimes
disturbed by the appearance of a body of large
toothed whales. Then every available boat is
launched, every Shetlander, or Orcadian, leaves
what he is doing to take an oar, and a vigorous
attempt is made to frighten the animals ashore.
The ca-ing, or bottlenosed whale, as the species
is called, seldom ventures beyond the Pentland
Firth.

Of the two other forms of toothed whales, the

dolphin is better known farther south, and is seen to most advantage when sporting under the prow of the sailing ship.

The porpoise (*Phocœna communis*) is at once the smallest of the cetacea, and by far the commonest species in the North Sea. Few who visit the seaside can be unacquainted with the shoals which rise one after the other to the surface, show their backs for a moment, and tumble back again into the depths.

The marine carnivore is the seal (*Phoca vitulina*). Though seeming to find the rocky west coast, with its islets, and sea lochs, more suitable to his habits, if not more productive of his prey; he is by no means uncommon off our shores. To secure the imprisoned fish, he sometimes wanders into the salmon-nets, where he is about as welcome as a shoal of dog-fish in the herring-nets; and employs his teeth to much the same effect.

He ascends the estuaries of our rivers, which are to the comparatively unindented east coast, what the sea lochs are to the riverless west. He is common in the Forth, and the Tay. As many as a dozen at a time may often be seen entering the St. Andrews Eden with the tide, and basking at the ebb beside the herring-gulls.

They remain with us all the year round, and seem to breed freely. In March or April, scenes which are associated with Greenland are brought near home; and the banks, exposed at low water, bear the same living freight as the Arctic ice-floes. Adults, with their young are dotted over the sand. In his daily walk a professor was surprised to see a strange creature, which turned out to be a seal, emerge from the water, and give birth to young. With a blow of his stick he secured it for the museum.[1]

There are few uncannie creatures in the North Sea, probably because the breezes are sufficiently strong, and fresh to blow away all delusions. However, stories, of a more or less veracious character, concerning the appearance of the mermaid, have been circulated, and attained a certain amount of credence, on the testimony of those, who had, at least, the merit of believing in the accuracy of their own observations. One of these composite creatures seems to haunt the northern islands, where it is so frequently and plainly seen, that it is possible to describe it with a considerable measure of detail. The upper, presumably the female half,

---

[1] The grey seal Halichaerus grypus is not uncommon, especially on the west coast.

is perpendicular, not prone, as in ordinary marine creatures. What could be more convincing?

This habit is so common with the seal, that one wonders how anyone, living at the seaside, could be unacquainted with it. Out on the Eden the other morning with the mussel-dredgers, I noticed a seal come to the surface, some fifty yards away. Evidently taken by surprise, the animal rose partly out of the water, as if to reconnoitre. While in this upright position, he had a startlingly human appearance.

The sea-serpent is undoubtedly absent; although I have seen, even in our unpromising waters, an atmospheric effect, or a tangle of drifting seaweed, or a row of tumbling porpoises, or a spouting whale, or a score of other things, which, acted on by the fertile imagination, supposed to be the birthright of skippers, were capable of giving rise to still greater wonders. I have just heard of the master of a fishing-boat here who saw one off Aberdeen.

Dr. Key observes of the basking shark (*Selache maxima*) that the large size, and habit of swimming at the surface, with its dorsal fin, and part of the back visible, and the jaw projecting out of the water, as it moves with open mouth in pursuit of prey, has suggested to ignorant credulity the idea of some large marine monster, which has received the name of sea-serpent.

As he grows to forty feet in length, this shark, even without the help of imagination, would make a respectable wonder; and, as at certain seasons he is gregarious, two in a row would be sufficient to cause the hair to stand on end. Happily he is not common off the east coast, and seldom attains his maximum size there.

Altogether the North Sea is an innocent place, except for its winds, which are sometimes blustering enough. There is not even a giant octopus to give rise to a blood-curdling story, and no enemy to life more deadly than the big brown jelly-fish (*Cyanea capillata*).

Only a few scattered sharks, belonging to different families; the big ones with sharp teeth, which they use on fishes, the little ones, and the flat ones, with blunt teeth used for crushing shells. The fiercest of them all can be trusted not to touch the man, who is wise enough not to touch them.

Only a few timid, and harmless mammals; all of which together add to the variety, and interest, without increasing the dangers of the deep.

VII

BAITS, AND SEA-FISHING

A S a sphere for the rod, the ocean has no great
attraction on this side of the Border. Trained
by the loch, and the stream-side; accustomed to
handle delicate tackle and dainty lures; Scotsmen
are disposed to regard the methods of deep-sea
fishing as coarse, and the skill required of inferior
quality. We have serious trout anglers, and still
more serious salmon anglers, men who take the
pleasures of life somewhat gravely; but to discover
a serious haddock angler, one would have to cross
the Tweed.

On the sea lochs of the west country, a little
mixed fishing may be indulged in, and a few
marine forms may find their way into the basket.
The grey mullet (*Mugil capito*), which comes up
with the tide, and lies basking in the pleasantly-
heated shallows until the ebb, may take the fly
8

meant for something else, and yield an interval
of exciting play.   But true sportsmen make a
distinction.

And, as for the east coast, with which I happen
to be tolerably well acquainted, I question if
there are half a dozen marine anglers between
Berwick and Duncansby Head; and if one of the
half-dozen could venture an opinion on the merits
of the Nottingham reel, or offer a guess as to the
meaning of "whiffing," or direct an inquirer where
to find a paternoster, except it might be in the
prayer-book.

Among those who are not afflicted with an
incurable tendency to *mal-de-mer*, there are a few
who venture out occasionally.   But, even such
would regard the refinements they love to lavish on
their favourite fresh water as thrown away on
unworthy objects; and would never think it worth
while perilling their souls with the conventional
white lies of trouting days, and achievements.   If
there are any who make a practice of doing that
sort of thing, they are not, as a rule, members of
any recognised fishing club, nor have they any
angling reputation to lose.

This stiffness may, or may not be excusable.   At
present I am mainly concerned in stating how

things are here; and in pointing out that the predominance of sea or river, probably the outcome of different surroundings and opportunities, is the main distinction between northern, and southern fishing.

The unoccupied field is doubtless large, and no one is disposed to question the assertion that it offers the prospect of a heavier basket than ordinarily falls to the stream fisher. Indeed, baskets are out of the question, unless they are clothes baskets.

The readiest bait is the lob-worm (*Arenaria piscatorum*), just as the readiest bait for fresh-water fish is the earth-worm. It was the first we dug in those days of innocence, when we too hung over the end of a pier, or stole out on the bosom of the Forth in the shoemaker's boat. The little cups in the sand, mark out its position, and the number of these cups, the multitude. A couple of quick motions with the graip, and he lies helpless, and is transferred to the rusty tin can.

It is the best lure for plaice—at least plaice-fishers prefer it. And, probably, as an all-round bait, it can be relied on to take more fish than any other.

Digging for lobs in the cool interspace between

ebb and flow, say in the twilight, or the later moonlight, when the ribbed seashore lies in long lines of light and shadow, short strands of living elastic are observed to strike first one end and then the other on the sand. And every here and there a thicker strand, equally lively, and about a foot long, appears among them. Those are the lesser and greater sand-eels, which have their use on the hook, but will probably find their way into the frying-pan.

After the lob, the more valued baits are found among the molluscs, chiefly the bivalves.

Encrusting the rocks at low-water level, to the depth often of half an inch, are multitudinous small mussels, so overcrowded, so chafed and knocked about by the surge, that they seldom come to anything. On the rough east coast, mussels grow to a marketable size only in the less exposed estuaries.

In these comparatively quiet stretches, the larvæ settle to the bottom as spat. This they often do on some high bank, covered only for half the tide. Exposed, for many hours every day, to sunshine or frost, and to an enforced abstinence from food, they may take six, or seven years to come to maturity. As a result of their slow growth, they secrete thick shells, and have a general dumpy appearance.

Or the spat may settle in some eddy, where they are constantly covered, sheltered, and fed.

The whole range of mussel culture lies between these extremes, and the secret of success consists in varying the less or more favourable conditions. After a time, the mussels are removed from the exposed banks to those of a lower level. And finally, they are placed in some quiet site in the bed of the stream, where they pass through the later stages with great rapidity.

When ready for use, they are brought to the surface with a formidable-looking rake, set with sharp teeth, and worked by a handle, varying in length according to the depth of the water. This rake is pulled through the scalp by a series of jerks, and raised to the surface with the clumps of mussels sticking between the prongs. I speak of the system pursued here; but should the handle prove unworkably long, as on the Tay, a dredge is dropped down instead.

The natural bed is still used in Scotland. The raised wattling, wherever adopted, has not as yet proved very successful. It is only a matter of time, until the silting of the estuary buries the whole structure. Very little more than the tops of the piles, represent an expensive experiment here.

To some scalps, which I recently visited, speedy destruction was threatened by the starfish, than which there are no greater pests in the sea. On each rakeful there were twenty, or thirty overfed fellows; and, as an evidence of their activity, nine out of ten of the shells were empty. Many of the raiders were at work. The stomach was exserted, so as to wrap the mussel round, and force it to open; when its contents were sucked out, and the shell dropped. The industry in this particular estuary is in danger of being palsied, and a remedy is being vainly sought for.

All this culture is meant for the behoof of those who do business in the great waters, and whose wives may be seen seated outside their doors, opening the bivalves with a skill which leaves no trace of mantle behind, and fixing the soft body with a few swift turns of the hook. The scene, in the windy street of the fishing village, is picturesque and savoury—from a distance.

As a good all-round bait, the mussel comes next to the "lob." It is the special bait for haddock, just as the lob is for plaice, with most of the North Sea fishermen.

The exception is the Firth of Forth. This, the biggest estuary by far on the east coast, and the

only one into which ships can run in all weathers,
is extremely rich in its molluscan life; especially in
bivalves. It has an immense oyster-bed, yielding

at one time a considerable
revenue to Newhaven, which greed, wastefulness,
and neglect have rendered barren, and useless. These
delicate morsels, however, are reserved for men.

More useful in the meantime is the common

scallop (*Pecten opercularis*). Although not fixing itself to the bottom, like the mussel (except in the earliest stages of its existence), but rather jerking about in the liveliest manner, by the opening, and shutting of the shell, it is still essentially gregarious, gathering in vast numbers into one place. A gigantic bed, twelve miles long; and a second smaller one, yield an almost exhaustless supply of bait to the fishermen of Newhaven, Prestonpans, Leith, and other places both on the Fife, and the Lothian shores.

The pecten differs from the mussel in being impatient of exposure, and more of a deep-water mollusc. Whereas, the range of the latter is, roughly speaking, from five fathoms shoreward, that of the former is from five fathoms seaward. The rake with its eighteen-feet handle is of no use in the Forth. Either the individual dredges for himself; or certain men make a living by dredging for the rest. There may be a danger of this haphazard mode of proceeding leading to exhaustion, similar to that of the oyster-beds; more especially as, unlike the mussel, the pecten is not amenable to artificial culture.

There is a little natural rivalry between the bait of the Forth, and that of the rest of the coast. The

former is declared by those who use it, not only
to be the more effective, but to have the further
advantage of sticking better to the hook. The
advantage, it is allowed, is less in summer than
winter, because of its inability to resist the warm
weather. The relative merits were put to the test
in a way which, if not conclusive, was at least
interesting. In certain experiments with the
various baits, made under similar conditions by the
Fishery Board's steam yacht, *The Garland*, pecten
was found to have a slight advantage in the
number of captures.

And, if the æsthetic is admissible in such a
prosaic matter, the pecten is your prettier bait.
There is no daintier shell in the sea, nor one more
delicately tinted. And, over the enclosed inhabitant,
Jeffreys waxes justly enthusiastic, pronouncing it
a study for a painter, "with its large and bright
pink ovary, and its mantle, studded on each side
with a row of brilliant eyelets, like dew-drops
glittering in the sun of a May morning."

Among the other Forth bivalves is the coarser
horse-mussel (*Mytilus modiolus*). This is a deep-sea
form, which grows to a much greater size; and,
although less generally used, is fairly effective
for cod.

The common cockle (*Cardium edule*) ranks about next to the common mussel. The range too is much the same, *i.e.* within six fathoms. And it shares the further advantage of not suffering from exposure on banks during the ebb; a hardihood which makes it available at all times, and renders it amenable to cultivation. The shell, however, is difficult to open; and scarcely worth the trouble, as the contents of two or three are needed to fill a hook. On the east coast it is more used for food than bait.

The clam (*Mya arenaria*) is similarly hardy with the cockle. The common limpet, used along with the mussel, makes an excellent spring bait for haddock.

When the breakers, driven before an easterly gale, tear up the sandbanks, and unceremoniously dislodge their inhabitants, they scatter an endless variety of bivalves along the coast; Mactra, Lutraria, Cyprina, Solen, Venus, and a host of others more distinguished for beauty than use. The quantity is endless, and would have served for weeks, had it only been given less imperially. The choice is embarrassing. The fisherman, his wife and children, are abroad with every available basket; which they bring back

brim full. Even such fleshy univalves as the whelk
(*Buccinum undatum*) are not overlooked.   But
the supply is too intermittent, even if the advantage was more assured, to entitle these to be
included in the list of available baits.

At a certain season, often when the sea is at
its calmest, and there is nothing in the nature
of the weather to account for it, a succession of
splashes, where the wavelets are lapping on the
sand, draw attention to the stranded cuttle-fish.
This probably occurs all along the east coast.   One
may count as many as a dozen thus rendered
helpless, in the space of half a mile.   If this
happened all the year round, and not for a few
weeks only, these might come to be reckoned in our
bait supply.   Cuttle-fish culture has been suggested ;
but, so long as other, and cheaper baits are available,
will scarcely be attempted.   At present, the cuttle-
fish used in the winter cod-fishing are bought
from the trawlers, at prices varying from two,
to three pounds a box.

An equally good bait for cod, and much less
expensive, when they are to be had, are the
common shore anemones, especially the dahlia
(*Tealia crassicornis*).   I find the fishermen here
call them " paps," a not inapt name, as they appear

during the ebb. So great is the run upon these, that they are in danger of becoming much less numerous. They seem to have the advantage over other lures of remaining alive on the hook for an indefinite time, of maintaining their bright appearance, and of having less attraction for those vermin, the starfish.

The question of bait is an anxious, and sometimes a vexed one; but its future is involved in a still wider issue. Should the trawler ever dispossess, and banish the boat, the mechanical mode of capture will supersede the lure. Bait will no longer be cultivated on any large scale. The various shore forms will cease to be of interest, unless to the naturalist or the lover of shells; except perhaps for the modest, very modest, demands of sport.

Perhaps one reason for our northern indifference to sea-fishing is, that some of the best fighting species are strangers to us. But, even were such to conquer their evident objection to our chiller water, the prejudice would remain. And, certainly, a northern angler would blush to the roots of his hair, if he were caught sitting on the point of a rock, or surprised in the following situation.

"When fishing from high piers, casting is anything but a convenient method of working the
bait. For bass, or pollack" (by the way pollack,
as has been pointed out, are not very common
on the east coast) "it is not, as a rule, necessary
to cast from such, because those fish are found
close to the piles of pier, jetty, and the like, and
many of them are caught by simply sinking and
drawing the bait through the water, or even
by walking up and down the pier and trailing
the bait."

The chief absentee among the sporting fishes,
is the bass. Though not unknown, he is so scarce
that he may be left out of account, as no one
would think of going out in search of him. He
bites eagerly when that way inclined, moves
about in shoals, and may be hooked up to 12
or 16 lbs. On the other hand, he is a foul
feeder, and not very good for food. Now, a strong
prejudice exists here against catching fish which
one cannot use, or even more than one can use.

"Though not particularly estimable at the table,
it ranks rather highly among the sportsman's
sea-fish, being plentiful, biting freely, and fighting gamely." Such is the certificate of character
attached to the bream.

These fish visit the North Sea in summer, and are occasionally taken on fishermen's lines in considerable numbers. The local Fife name is Jerusalem haddock, doubtless given because of the mark of Peter's thumb on the shoulder. They mainly frequent the deeper water; the comparatively flat east coast, not being favourable to an inshore life. Their very presence is unknown to the average amateur angler.

Plenty of sport, of an inferior kind, may be got out of lythe and saithe. The latter abounds, is to be met with everywhere, is easily hooked, gives sufficient play, and grows to a considerable size. But he does not appear in the shop windows, is not served up·on dinner-tables, is not such as the successful angler cares to eat himself, or impose upon his friends. And, when the excitement is over, one cannot help feeling a little nasty as he surveys the 40, or 50 lbs. lying in the bottom of his boat, for which he has no conceivable use.

An ordinary Scotsman—of course there are exceptions to all rules—would as soon think of eating a snake as a conger-eel; and, if, by any chance he hooked the latter in trying for some other fish, he would be extremely glad to sever the connection at the expense of his tackle. Any instructions as to how these monsters are secured is thrown away upon him. I have seen an angler hook a common eel in a flooded stream, and I have found his face a study in the expression of the emotions.

The true sporting fish in Scots waters are, among the round species, cod, with occasional whiting and haddock, the last mainly taken with bait. Some would include a humbler fish. "Herring have, I believe, a bright future before them, as well as a respectable past; for there are many instances on record of these fish taking the fly in a most complacent, and entirely satisfactory manner." A friend, fishing one of the sea lochs of the west coast, had the pleasure of hooking a few, but I am not aware that he regarded it as sport. Really, we must draw the line somewhere.

In the months of May, and June, the most delightful time of the year, "jigging" for herring is largely practised by the fishermen; but may be adopted by amateurs, alike to their pleasure, and profit. The arrangement consists of a string, with seven knitting wires, at intervals of fourteen inches. From each end of the knitting wires, a bare hook is suspended—fourteen in all. The two-pound sinker attached to the end, is egg-shaped, to lessen the friction in both directions.

The aim is to reach the fishing-ground about sunset, the most delightful time of the day. The line is dropped over, and jerked up and down—hence "jigging." Attracted by the glitter as of sand-eels, the herring take the hooks eagerly, generally filling the whole. Two lines are worked at the same time, so that one can always be in the water to keep the fish near the boat. As many as three or four hundred herring may be thus taken. The sport only lasts for about an hour. When the phosphorescence comes out on the surface in the deepening twilight, it is time to stop.

The jigger is now exchanged for a hand-line. The herring, or some of them, are used for bait, and the short summer night is occupied in fishing for coal-fish, and cod.

When the phosphorescence disappears from the water in the growing morning light, the jigger is once more dropped overboard, and worked till the rising of the sun. Then, the fishermen sail away to their distant crab-traps, which they empty of their living freight, and bait again with the coal-fish, and, if need be, the cod,—a curious three-cornered arrangement.

Most people here trust to the fisherman to supply the tackle along with the boat. Half a dozen

squares of wood, of the diminutive Oxford frame
pattern, such as some boys use for their kites, are
thrown in. A quarter of a mile off the rocks, the
anchor, or its substitute, is let down. Line is paid
out till the lead strikes the bottom, and then pulled
an arm's stretch back again, to clear the hooks
from the ground.

Usually within a few minutes, the first sharp tug
is felt, and met by an equally sharp check in the
opposite direction. Hand under hand, if possible,
without a pause, the moist, and slippery line is
drawn in. At length a white gleam appears dart-
ing through the deep. A lift clear of the side, and
a brown-backed cod of 5 lbs. is on board.

A few tugs intervene before another is hooked,
and the hauling in begins afresh. This time the
fish has a deeper shadow on its back. It is a
saithe.

Then comes a pull on the finger, so gentle, that
the expectant angler waits for a second. The line
comes away without effort, and a mask of a fish,
big-headed, and tapering swiftly away to the tail,
is brought on board. This is the grey gurnard
(*Trigla gurnardus*), whose acquaintance will be
made, only too frequently, within the next two
hours. He is really good for food, but there is a

9

prejudice against him, probably because of his forbidding appearance. The tough white skin of his
breast will serve as a bait for mackerel.

A shift to the sandy bottom, opposite the bathing-
coaches, will yield flat fish,—dab, and plaice; and
another to the bank, two miles farther out, whose
position is marked by the steeple over the distant
woods, will add haddock, with possibly a larger cod.

Pleasanter than thus riding at anchor, and less
likely to be productive of disagreeable consequences,
is sailing with a gentle breeze, say at two knots an
hour, fishing meanwhile. The varieties of fly are
almost as indefinite as those of bait; and, if one
chooses to open his salmon book, he will get an
offer for any of the named varieties. A less expensive lure will probably be equally attractive.

A bright spoon bait, and half a dozen white
feathered hooks, attached to a common line, and
dropped over the stern, is often productive of large
fish; sometimes so large, as I had reason to know the
other day, that they run everything out, and break
away, before there is time to slacken speed. Two
hooks, tied together back to back, are sometimes
used.

If this be whiffing, then whiffing is not without
its excitement.

And, when one has sailed across the bay, and is miles from home, and the soft warm twilight of summer has shaded the sea, and softened the outlines of the coast, he can put up the four pieces of his rough bamboo rod, and tie some white-winged flies of a smaller size to a thinner cord, and cast away to the leeward. It will go hard if he does not get in among the poddles, as the young of the saithe and lythe are called; and, when once in, his fun will be fast, and furious. If one takes the hook, it only attracts the others, till half a dozen may be pulled out in a row. It seems a great shame, seeing that not one of the fifty, or sixty caught will ever be used.

In this crude state, I am afraid, the art remains in Scotland. It may, as an enthusiast puts it, have a great future. It is undoubtedly susceptible of greater delicacy of method, seeing that it could scarcely be rougher. If one really wants larger numbers, he will do well to use finer tackle; but I never yet heard anyone complain that he had not enough of the kind. In any case, in the interests of true sport, I should certainly recommend the fisher to keep to edible forms.

Whether the northern angler will catch the infection is, I think, more than doubtful. Many of our clear lochs swarm with perch, or fresh-water bass, lying among the submerged grasses, which might be had by the cart-load. And yet no one, above sixteen years of age, thinks them worth rod and line.

Even pike, of from 15 to 30 lbs., are left unmolested in their weedy haunts; and, should occasion arise to thin them out, men are sent down with a net. After pike, and perch fishing are seriously pursued, bass, and saithe may follow.

Sea - fishing has a charm, lent by its very association with the blue water, and the fresh breeze, and the magic coast-line; and serves a useful purpose, as a pleasant break on the intolerable monotony of a month in a watering-place. But, neither of these entitles it to a place in the first rank of sport.

VIII

SHETLAND—MIDSUMMER

I FACED northward towards Shetland for a holiday, to be devoted half to science, and half to sport. The science was to be done in a leisurely, and unexacting fashion; not incompatible with a considerable amount of freedom, and enjoyment; the sport was to be kept within certain limits of an exceedingly elastic description.

The serious object, if such it could be called, was to compare the life of the warmer currents, which break the northern termination of the land into so many islets, with that of the chiller waters of the North Sea. The only scientific encumbrance was a tow-net. Now, as this sort of work means boating, to which a net dragged through unknown waters only adds a zest, science, and sport mingled together in the most friendly fashion imaginable. And, the rough expedients for the examination of the catch, in wild and out - of - the - way quarters, added a touch of the ludicrous. The impression of the small deep-set window, in a certain un-mortared and tumble-down cabin, into which two heads were often thrust, in the vain attempt to see if there was anything fresh, will not soon fade.

The common tern drops more and more behind; until only the Arctic form is left. This is Shet-land's southern migrant. The northern migrants, which come in the winter, are away within the Arctic Circle; the great glaucous and Iceland gulls following the whalers, and that shy snowbird the ivory gull, farther north, keeping company with the walrus.

The common gulls of the North Sea remain undiminished in numbers; while the rarer species, such as the lesser, and greater black-backed gulls, become common. If there is any advantage, where all abound, then the local kittiwake be-comes the dominant form. In incredible numbers it occupies every available ledge of these stupend-ous breeding-places.

A like increase is observable in other sea birds. We learned that, in the latitude of the Firth of Forth, we were simply among the outliers, or outswimmers of the divers, guillemots, and razorbills, and were for the first time coming in contact with the main body. Puffins, by dozens, bob on the waves, and cormorants, not one but many, occupy the exposed rocklets, or bend their horizontal flight across the water.

Among the forms less familiar to us in the south are the shearwater, the fulmar petrel, and the great skua. Than the last of these three, there is no native nearer, or in greater danger of extinction.

The more interesting land birds are those one would expect in a treeless scene, i.e., such as do not perch, and build on the ground. While the rose linnet, a bush or furze haunter, though reported, is either absent or very rare, the mountain linnet, or twite, with which the former is probably confounded, is common. I asked some boys to show me the eggs they were in the habit of getting, and found them to be mainly those of the latter species.

The wren greets one everywhere, and it is perfectly delightful, and homelike to see him popping out from the dry stone dykes which surround the crofts, or to hear his loud pipe amid the wildest rocky scenes.

Happily for the gaiety of the place, the lark abounds. The meadow-pipit is numerous; the rock-pipit still more so. The former is named the hill sparrow, from his more inland resorts; the latter the tang sparrow, from his preference for the coast; and thus the two are happily, and correctly distinguished by their popular names. The wheatear, locally known as "the steinkle," arrives in April, and, throughout the summer, his click and flash are familiar.

All these are musical, either in reality, or by courtesy. Farther south, we are not in the habit of talking about the vocal powers of the wheatear. Perhaps we are too particular, as he has a loud, and not unpleasant song, which only suffers by comparison. And yet, he is included among the Shetland singers.

Sparrow, and starling are found wherever there is a croft. The yellow-hammer, and the robin, which frequently build on the ground, are there in small numbers. The two wagtails are present, but rare. The hedge-warbler, and the song-thrush are only seen at intervals, driven it may be by storm, and seldom, or never breeding.

The blackbird drops down in autumn, and spring; so too does the mountain blackbird. Many other species use the islands as a convenient resting-place, in their double passage to and from their nesting haunts.

The hawks, also, are those which frequent open places, or bare sea-cliffs. The sea-eagle is a common form. The peregrine falcon sweeps from his perch in pursuit of the guillemot, or the razorbill. His miniature, the merlin falcon, pursues, with a similar flight, the lark, and the meadow-pipit. Absent in winter, the kestrel is found in summer hovering over the shivering rodent. Greenland falcon, and hen harrier are not uncommon. The goshawk is rare.

Among the night-hawks, the short-eared owl is common in spring, and autumn. The long-eared owl is rare. The barn, and wood owls probably absent. The snowy owl is occasionally seen; possibly driven over by storm, although he has been supposed to breed.

The hooded - crow abounds, since there is no gamekeeper to hold him in check. The carrion-crow is rare; thus reversing the order for the south. The raven is one of the common sights, and his croak one of the common sounds. Probably, he too profits by the absence of persecution. He frequently attacks the hens, so largely kept by the crofters.

Most interesting of all, perhaps, are the rock-pigeons, the reputed source of our breeds. Not that they are unknown round the coast of Scotland; but here they share the sea ledges with the kittiwake, and, in their flight across the land, take the place of our common pigeons.

Shooting was prohibited as it was close time, and we had omitted to provide ourselves with a special licence as naturalists. Personally, I was not sorry, as I object to any more stuffing than is absolutely necessary. A stuffed bird always makes me melancholy, affecting me very much as an Egyptian mummy does. Probably, the resemblance to life is equally close. Therefore, my objection to enter a natural history museum.

In ignorance of the prohibition, one of our number shot a cormorant, which was diving about twenty yards from the shore. Unwilling to lose his reward, he endeavoured to swim out to it. Between him and the object, were certain long filaments, waving freely in the water. As he struck out, these wrapped themselves round his body. Each stroke brought him within reach of new filaments, while the old retained their grasp. Alone, and beyond the reach of help, he forgot the bird, and was only concerned about reaching the shore. By the act of turning, he only wrapped himself so much the more tightly in the coils; and, it was by efforts well-nigh superhuman, such as one puts forth when it becomes a case of do or die, that he broke loose.

Till now, my acquaintance with *Chorda filum* was confined to rock pools, where it seldom grows more than a few feet in length. But it seems that it reaches the upper air, however many fathoms deep the water may be; and there creeps fathoms more along the surface.

Many of the voes of Shetland, out of the scour of the tides, are quite choked by it. When rowing, the filaments wrapped themselves around the oars, as they did round the body of my friend. Their local name is "witches' hair"; and this was the weird meaning thereof.

Of game, Shetland has none; for probably none would live, not even a hare. Such haunters of the areas of cultivation, such lovers of warm well-drained fields as partridge, cannot reasonably be looked for on this rude, and untamable soil. One attempt at introduction has already failed, and no other need be made, or has a chance of succeeding.

Of late years, enthusiasts have tried more than once to naturalise grouse; but the experiments appear to have been conducted carelessly, and have so far, proved fruitless. Mr. Harold Raeburn considers that there is no reason why grouse should not become perfectly well established in Shetland, if sufficient trouble were bestowed on the effort. The birds would have to face some drawbacks here, as elsewhere, *e.g.* "the dense population, the wet and stormy springs, and the sinful custom of burning heather all over, instead of in suitable strips."

My experience of Shetland is of a land, sufficiently thickly populated no doubt for the extreme poverty of the soil, but utterly lifeless in the main, for want of a sign of human habitation; of a land too, so sodden, even up the hillsides, that my feet were seldom dry.

If, even in Argyllshire, the number of grouse is considerably smaller than in Perthshire, for no want of heather, but simply because of the addling of the eggs, one is led to judge that in Shetland not a single egg would escape.

The number of Royston crows is an objection which could be more easily reckoned with, than the irreconcilable soil, and the almost constant mist, and drip.

Rabbits abound; but the natives seem to have a prejudice, common in such outlying districts, but which I could not trace to its source, against using them as food.   Under the circumstances, they would probably multiply to an inconvenient degree, did not nature apply her usual checks. The increase of stoats keeps pace with the increase of their prey.

The centre of our activity was the parish of Northmaven, at the north-west corner of the Mainland. This is the highlands of Shetland, rising above the curse, and exposing to the light of day its red or silvery summits of gneiss, or granite.

We sought the voes in all directions, walking often great distances to reach them, and embarking in such leaky craft as were available. And each had some fresh surprise in its surroundings, or its shape, or in the islands floating in its centre. And all teemed with life, which varied within certain narrow limits, according as the inlet was connected with the Atlantic, or the North Sea.

There is always that finishing touch to every striking natural scene, if one has only the good taste to find it out. And, in the case of the voes, this is provided by the omnipresent tystie, or black guillemot—black with a patch of white on the wing. This is the Shetlander's darling, or good spirit, as well it may be.

There are inland lochs as well as voes, but these

are for the most part dark, from a dense bottom
accumulation of peat. No lively tystie is there,
only an occasional cormorant. Boats are unknown
luxuries; so one merry day, by the aid of some
Shetlanders, we trundled one up the hillsides,
across the plateau, and down into the basin that
seemed most promising. Never before had line, or
human shadow been cast across its surface, except
from the shore.

> We were the first, that ever burst,
> Into that silent sea.

The trout are, perhaps, not so plentiful as in
the Orkney lakes, beside which I had tented on
a previous year; but they make up in size for
what they lack in number; and are sufficiently
uninitiated in lures to rise eagerly; and lively,
when hooked, to afford excellent sport. They are
dark in colour, except where they lie on patches
of clear gravelly bottom near the edge. While
some fished, others lit a fire to cook the spoil;
and so we spent several gipsy days.

But, the appetite for trout has its limits; and
fishing ceases to be sport, just in proportion to
its success, when one is compelled to eat all
he catches, whether he likes or no. This seems
to be the weakness of fishing in out-of-the-way

places. What are half a dozen people to do with 25 lbs. weight of trout, easily obtainable every day, except to let three-quarters of them go to waste. Unable to invent any excuse for further destruction, one begins to ask himself whether he is justified in going on. An ingenious member of the group, by the aid of some oak-chips, tried his hand at a species of curing, with indifferent success; and, at length, not without a sigh, we resigned ourselves to fishing for the larder.

Some of the lakes are in an intermediate condition, isolated when the tide is back, and flushed twice a day, as it flows. The narrow neck, through which the sea finds entrance from the voe, may be compared to a river, contracted into less than a hundred yards. Without moving from the spot, one can command a constant succession of running fish. And the silvery sheen of the great sea-trout is quite delightful, after the dark hues of the lake trout.

Sunk sea-lines had also their attraction, and never can one who has experienced it forget the musical dip of the oars in the fresh morning, or in the magic northern summer night; or the successive splashes of the great flat fish, as, one by one, they were hauled to the surface.

Small cows, probably allied to the Celtic short-horn, gave a scanty supply of milk, which was purveyed in ordinary glass bottles.

The sheep are the most delightful little creatures in the world—very small, fragile, and thin-legged. Exposed to no selection, either natural or human, there is the usual tendency to vary, especially in colour; and very beautiful they look. There were black sheep and white sheep, in almost equal proportions; but these are but the web and woof out of which the most varied effects are wrought. The threads, playing freely in nature's loom, produce sur-prises which would have baffled ingenuity, or the wiles of Jacob. "And the flocks conceived before the rods, and brought forth cattle ring-straked, speckled, and spotted." The black and white appear in patches, and circles and streaks here, there, and everywhere; and in no two alike. Occasionally, one is born brown; and because of its rarity, and for no other reason, is of considerable value. The natural colours are sufficient for the ordinary purposes to which the Shetland wool is applied, although in some districts vegetable dyes are used.

Charmed with their beauty, and fairy size,

a friend chose the most pigmy, and delightfully
coloured of the lot, to take home as a playmate
for his children.   But, on the richer feeding of

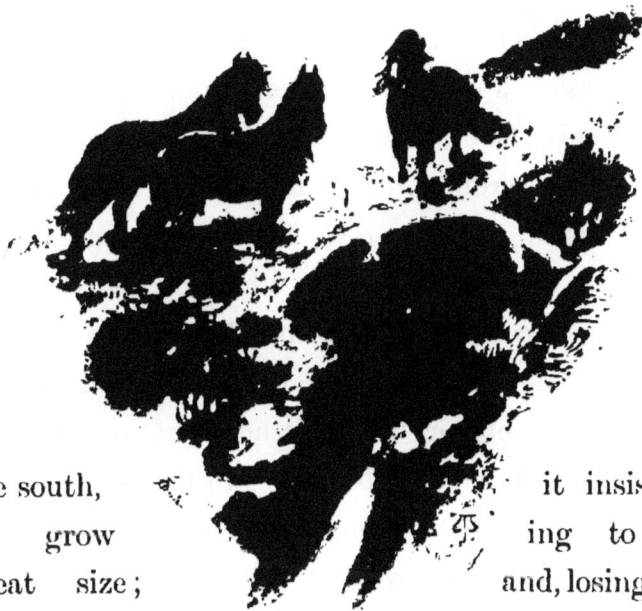

the south,   it insisted
on   grow   ing   to   a
great   size ;   and, losing all
resemblance to   its   former   self,
became   posit   ively ugly.

The ponies appear by twos and threes, grazing
upon the dark hillsides, leading down to the
black peaty lakes, often very far from human
habitation.   They seem to be left to themselves

10

until required for the market, and to lead a perfectly wild life, being neither fenced, nor fed, nor
otherwise cared for. Some grow comparatively
big, others remain fairy-like, as if designed to
bear Titania's chariot through the glimmering
night. The former are kept at home, and used
for the transport of peat, and other purposes.
The latter are exported; and once brought fancy
prices. A one season's foal, some two or three
years ago, was sold for ten pounds, a perfect
fortune in Shetland: probably the only money
the inland crofter saw; unless he went to the
whale fishing, and even that is failing him.

The Americans were the chief purchasers. All
attempts to breed them in America, and such
seem to have been made, failed. The foals invariably grew bigger than either sire or dam; a
well-merited rebuke to the greed which would
rob a poor man of his one source of income. This
would seem to show that the small size of the
Shetland pony, probably also of the sheep, is
largely owing to the hard life it has to lead,
and its coarse, scant fare. It is one of the most
striking examples in the animal world, of the
influence of environment as distinguished from
natural selection.

The evenings are long. There is practically no darkness. Our own larks sing late and early; not stopping till eleven, and beginning again at two. But the Shetland larks are said to fill up even the short interval, and sing all night. The farther north, the more lingering the twilight. Night overtakes the morning, and the lips of dark and fair meet.

The latest development of golf was a match at midnight, in which the competitors seem to have experienced no difficulty in finding their balls. Happier they than many who play by day.

The natives, when a distinct type, become interesting, and fall within the limits of natural history. Among the hills one is always arrested by a pure, trousered Celt, as distinguished from a kilted, and mongrel Sassenach; or by the small, dark, long-headed type not uncommon on the west coast.

The Norse strain is comparatively pure, although not without an admixture of suspicious names, and complexions; certainly it is much purer than in Orkney. For one thing there is little to tempt settlers. Any London *debutante* might have envied the girls, who sat at some of the cottage doors, their transparent skin, and sunny, or golden hair; and the picture would have been enough to bring the picturesque spinning-wheel once more into fashion. Alas! for a little romance, which never got beyond the first act.

The men are loosely put together, probably not physically strong, certainly over-given to tea-drinking, a form of dissipation whose innocence is only apparent; but courteous and helpful withal. The curse of greed, and servility, which shadows one through the haunts of tourists, has not reached those outlying districts, or blighted the manhood of the simple natives. With little to occupy their lives, they are always ready to oblige; and the difficulty is not to satisfy them, but to get them to take any return. As a result of our intercourse, we came away with the pleasant feeling that we should like to meet with them again.

IX

LOCH LEVEN AND LOCH TAY

I HAVE caught flounders, in one's flounder days, amid the brackish waters, where the overflow of Loch Leven finds its way into the sea. Then I remember to have heard from the lips of old men, who may have been partakers in the fray, tales of new run salmon, when as many as fifty were enmeshed in one sweep of the net.

In Sir Robert Sibbald's *History of Kinross-shire*, we read that salmon, and even flounders find their way up to the loch. This refers to a hundred and seventy years ago, but must have persisted to a much later date; probably, through the first quarter of the present century.

For several miles from its mouth, the Leven is now so polluted that nothing with fewer lives than an eel could survive; an instance of what has been permitted to happen in the case of far too many

149

of our Scots streams. Beyond that, the current becomes sufficiently pure for life; and I have a vivid recollection of spring days with the rod, at the meeting of the waters where two tributaries enter.

Wiser counsels now prevail; and halting efforts are being made to clear up the impure interval separating pure stream from pure sea. But, whether the salmon will return to close the reign of the trout; or, whether this is a consummation devoutly to be wished, may well be questioned. The larger fish could scarcely enter, without disturbing the balance of life, and changing the present complexion of sport. Now, we have plenty of good salmon lochs already. This is the one good trouting loch in Scotland.

Some distance from the loch, the stream becomes straight, almost as the flight of an arrow; and flows between banks which are manifestly artificial. Thereby hangs a tale. In December 1830, the shallow, winding, natural tail-stream, was deepened and altered, to give more water to the mills on its course. The area of the loch was considerably lessened, to the benefit of the surrounding pro-prietors.

Many a time have I sat opposite the low Island of St. Serfs, dangling my legs when they were shorter, over the embankment, where the swift glancing water enters the sluice, drinking in im-

pressions,
meanwhile, of
the mild scene,
with its one bold
feature, the Lomonds.
I have fished it since,
perhaps to more purpose, but have never managed
to extract the same enjoyment; at least the colours
are not so fast. I have swum out to St. Serfs,
or boated to the better known isle. Loch Leven and
I are old friends.

The scene has long been a resort of the curious,
the patriotic, and the sentimental.   Its attractions
were discovered long years ago.   Each of the two
islands, once farther from the shore, has pre-eminent
claims on a separate order of devotees.   The in-
cidents of the one are just sufficiently far off to
have their harsher features veiled in the distance,
which lends enchantment to the historic retrospect.
Perhaps it owes not a little of its undying
attractions to the romantic genius of Scott; and,
certainly, few who go are unacquainted with *The
Abbot*.   The story of the other, as it was enacted
in the dimmer past, needs the deciphering of the
antiquary, and is thus of more restricted interest.

Dreamy-eyed lovers of a vanished beauty, rowed
across to muse an hour amid the ruins of Mary's
castle prison; if by moonlight so much the better.
With no less fervour beetle-browed Monkbarns
turned the prow toward St. Serfs.   Time was
when these two groups formed the majority; now
they are swallowed up, and scarcely noticeable there.

The associations of the loch are no longer with
hermitage, or prison.   It is now the delightful
resort of the pleasure-seeker.   The new enthusiast
arrives at Kinross station with a large wicker-
basket, and several rods, while a fat hook-book

bulges out his coat - pocket.   Probably · he has
engaged a bed by letter, or telegraph; but, if not,
he hastens to the hotel to supply the omission.
With the comfortable sense of shelter, he goes
out to have a look at the scene of action, and
to see about a boat.   His eagerness may suffer a
check, and his patience be put to the test.   Scarce
a summer day passes without the incursions of a
club or clubs, who may, weeks before, have secured
every plank for their annual competition.

When at length his turn comes, he drops a
minnow over the stern, or maybe his boatman does
it for him; in hope of a big one, while on the way
across.   Nor is he disappointed.   The chirp of the
wheel catches a practised ear; the oars cease to
play; and the net is got ready for action.   In his
awkwardness, he lets him run under the keel.   The
tackle holds, and first blood,—"two pounds if an
ounce,"—is drawn; or it may be even a heavier
weight; in which case the duller drag on the line
announces a pike.

They near the shallows, where the experience
of the boatmen tells them that the trout are lying.

The very first cast brings a rise, which takes the
fisher unaware, and comes to nothing.   "Canny
neist time," and that next time is not long in

154 WILD LIFE OF SCOTLAND

coming. There are two rods on board. Why then should one lie idle, when trout are so keen, and one man can keep the broadside to the shore? The second rod is put up, the other boatman begins to cast, and between them they have a merry time. It is a lively loch, when the rise is on. The stranger gets credit for the entire basket.

There is no limit to the time one can fish on a summer night; no limit to the delight of the thing. The grateful coolness, the shaded light, for it is not dark; the ripple of water, all weave themselves into an unique experience. As they return, the outlines of the castle are softened into dreaminess, awakening pensive and gentle thoughts of her, who must often have looked out on the scene on such a night.

The Loch Leven trout is not so deep in proportion to its length as the burn trout. The red flesh suggests molluscan diet. It is found in various lochs, in the south of Scotland, and north of England, presumably where the conditions of life and food-supply are similar. Beyond that range it does not extend; it is exclusively British.

The origin is obscure. Either the brown trout has been thus modified by long residence in the lake, or the sea-trout has been shut in, and adapted

itself to a fresh-water life.   It crosses freely with
both ; its affinities seem to be with the latter.

The loch owes its pre-eminence to the high average
in size and weight, compared with larger, and more
promising sheets of water.   The exception is to
hook anything much under a pound.   No fisher
baskets fry that are reckoned by the number.   Upwards of twenty thousand such trout are captured
every year.   This is a tradition of the water, and
suggests some cause, which I shall discuss elsewhere.

To meet the waste, as many as three hundred
thousand fry are sometimes introduced at one time.
And a sufficient supply is always on hand to satisfy
the demands of dwellers by other lochs, and streams,
who wish to have such an enviable commodity
nearer home.   In many cases in which they are
sent to a distance, and introduced into new waters,
the purchase can scarcely be worth the trouble, and
the money.   There is the risk of the new-comers,
after a few years, becoming undistinguishable from
the natives, and degenerating into very common
trout indeed.

It will always remain a question whether those
artificially-reared fry, when turned out to fend for
themselves, will grow into as vigorous adults as
those deposited by the spawning fish, in the feeders

of the lake, and flushed with the natural currents. And, after all, it seems certain, in the case of rivers and lochs, competition waters being doubtful exceptions, that no amount of legitimate fishing will seriously lessen the number, or do more than help to increase the size of those that are left.

Loch Leven is a lowland sheet of water, simply gathered into a hollow on the surface deposit.

The Highland loch lies in a clear stony basin, scooped out for it in the solid rock. The scratching of the great ice plough, which did the work as it grated along, and bit its way down, is still in some places not quite erased, and may be seen in Loch Tay. It is too deep to be drained, and, if laid bare, would probably prove too barren to be of any use. Happily, too, it is of much more value as a picturesque element in the scene, and a substantial addition to the sporting attractions of the district; otherwise there is no telling what might happen. Whether the tail-stream will fret the channel so deep as to run it all away, and expose the bare bed, as tail-streams have done for a thousand other lochs, we may leave to the future, and rejoice meantime in the romantic charm.

Just as no Scots burn is without its resident trout, so no Scots stream, with its lakes, unless

rendered inaccessible by natural barriers, or shut
off from the reinvigorating sea by contaminated
water, is without the migratory species, chief of
which is the salmon.   Common trout, lake-trout,
sea-trout, and salmon, meet and mingle in its
common waters.   The Clyde is the only stream
of the first magnitude which is salmonless.

Taking advantage of the attraction thus offered,
hotels have sprung up round the shores of the
lochs, and hold stretches for the behoof of their
visitors.   Upwards of fifty such hotels have preserved waters.   Probably, they would also offer a
day's shooting, if hillsides were as cheap as lochs.
In that case a mortuary might prove convenient.

Arrived at the lochside, the tourist embarks for
a day's sport.   Before the first of May, his boat costs
him twenty-five shillings a day, because the fishing
is better then, especially before the nets are put on.
After that date the charge is reduced to a pound.
He is not expected to provide lunch for the boatmen.   But, should he be kind enough to do so, he
is most earnestly requested not to give more than
a soda-water bottleful between the two.   The consequences of exceeding this fairly liberal allowance
are not stated.   He may never have had a rod in
his hand before, except to fish for poddles, or, it

may be, bass, over the end of a southern pier; but that does not matter. The boatmen are quite capable of doing everything for him, except actually putting the fish on the hook; and will prove the most obliging of men, especially if he do not adhere too strictly to the lunch limit. To troll is as easy as to haul in a boat-line — which his cast often resembles in strength, if not in thickness—and the largest fish are got by trolling.

Even fly-fishing is brought within his reach. A stiff breeze will carry out the line without any act of his, and, at the worst, the shortest cast is often the most successful. In such open water there is no danger of hooking the bank behind, or the branches above, or the boulder in the centre, or any other foreign body, except it may be the boat.

I spent a most uncomfortable forenoon on this same Tay, dodging a hook which seemed to appear on both sides, and all round my head, at once. At length the lethal weapon entered the lobe of the boatman's boy's ear; and I was selfish enough to congratulate myself that it was no worse, that is because it was not my ear. It was the first and only blood that day, and we were all heartily glad when the time arrived to go on shore.

It is one of the commonest experiences of those

who go down to the loch in boats, that a trout, a
ferox, even a salmon, will sometimes have the bad
taste to prefer the hook of a duffer to that of an
expert, to the intense disgust of the latter. If he
is not very nervous, sometimes whether he is or no,
he will land it triumphantly; although, as between
the fish and himself, he is the more astonished of the
twain. All the while he is as little of a fisher as
of an astronomer; or as his wife, who hooks the
next salmon, and lands, or boats it too.

And, should the fish be dour, or the weather
hopeless, and every wile taught by experience fail;
there are consolations in store even for the unsuccessful. These boatmen have learned to say
smooth things, more's the pity; and administer the
flattery with a skill begot of frequent practice,
and a fine discrimination worthy of accomplished
physiognomists. Although they might save themselves the trouble, as I never knew a golfer, or an
angler, who could not swallow a pretty strong dose.

It is reported of one that, returning from an
excursion, whose non-success even his skill could
neither prevent nor conceal, and unable to think
of anything else, he said, with an inscrutable expression of countenance, meanwhile: "That was a
fine fish you didn't catch!"

The somewhat enigmatical compliment may have been meant to suggest some problematical rise, which, had it come to anything, would have produced astonishing results.

In this case, the tail-stream is but a continuation of the lake, alike in its picturesqueness, and in its capabilities of sport. Indeed, there are those who prefer the romantic and changing charm of the running water, to the still beauty of the lake; and take more delight in a cast from the green bank, over the shaded pool, than on the opener water. There can be few rivers whose course is better worth following than that of the Tay from Kenmore, to its junction with the Tummel; thence through the wooded district leading by Birnam, and across Strathmore, till it meets the advancing tide in the neighbourhood of Perth.

Hotels dot the course, every here and there, which lease the stretch in their neighbourhood. And, doubtless, fishing is made as easy for the inexpert as the changed circumstances will allow. Boats are to be had, where they are in demand. And, there be stepping-stones in the salmon pools, and other aids to such as mistake herons for eagles, and respectfully, but firmly decline to wet their feet.

X

BY THE LOCH-SIDE

ONE of the brightest experiences of the year
is my summer walk through some portion
of the Highlands, without any definite purpose
except to take in new impressions and pictures,
or to freshen old ones; and without any fixed
finishing to the day's journey. A programme
terminating each night in some comfortable hotel
would spoil it all. There is to me a delightful
freedom in a planless wander; and a no less
delightful uncertainty, and variety in the shelters
for the night.

Toward the end of July, I started from Comrie,
to skirt the lakes between Loch Earn, and Loch
Katrine, including Balquhidder and Strathyre. This
district, studded as it is with lochs, and, gathering
within a little space all that is most charming in
the Highlands, can never be gone over too often.

11

Pairs of unprotected female tourists were on the road, evidently touched with a like Bohemian mood, and trusting to chance for lodging when night fell. This struck me as a pleasing testimony to the security of the times; and a comparatively innocent manifestation of a modern spirit, with many more unpleasant developments. But it scarcely belongs to the wild life of Scotland.

Redstarts abounded, especially in the neighbourhood of the few scattered cottages. I was the more struck with this, having just come from a country-side which they never visit. After all, only comparatively few species are universally distributed. The majority of so-called common birds are so uncommon twenty miles away, as to be practically unknown. The lists of local observers are mainly useful for comparison.

The redstart, like the wheatear, is an exception to the rule that a bird is usually sober coloured when flying away, as if to conceal it from a pursuing enemy. The majority of species display such attractions as they possess, on their approach.

A yellow-hammer was sitting on a stone dyke, trilling out his simple and yet not unpleasant lay; for he is a very late, if not the very latest of our spring singers. A pair of reed-buntings were moving, in their lively way among the bushes by the lake-side.

Both are equally beautiful, indeed all our buntings are, except the corn-bunting, which is the clumsiest of our small birds. Here the brush has been run over a different part. The yellow head of the one is changed to black in the other.

If the bright tail has anything to do with the guidance of the young, both sexes will possess it, as the female must need it more than the male; and this is the case alike with the white of the wheatear, and the red of the redstart.

If the conspicuous head is an adornment which would be unsafe in the sitting bird, only the male should have it. And the black, and yellow are confined to the male buntings.

These things, though not conclusive, seem to point in a certain direction.

A merlin, on the ground ahead, allowed me to come within a few yards. When, at length, he rose, he seemed to be attached to an object larger than himself, and much too heavy to carry beyond a short distance. As often as I approached, he retreated; never succeeding in rising above a foot, or lifting his burden free from the ground. Under the impression that he was trapped, I hurried forward.

When he could no longer avoid me, he made a supreme effort; but, too eager to watch his direction, dropped into the lake. And, not till he was in danger of being drowned, would he consent to loose his hold.

When fished ashore, the object proved to be a large missel-thrush. He attempts bigger prey than that, but what struck me was his determination not to be driven away. I had heard that the bird was the maximum of spirit in the minimum of size; and here was an illustration. It was that indomitable pluck, together with a tractable disposition, that made this little falcon such an excellent hawker.

The kestrel was hovering about. By the way, this hawk has been known to strike the pursuing weasel instead of the fleeing mouse, and suffer the usual penalty. But long experience should have taught the simple distinction, and this occurrence must be rare.

The rough strip of wood, on either side, was alive with young birds, which the old ones were fully occupied in feeding. Now is the time when the hawk can lead an idle life, obtaining abundance of food in the plump little fledglings, unable to fly any great distance. The immense destruction taking place helps to explain the fact, that, though so many eggs are hatched, the number of birds in a district scarcely sensibly increases.

Mire snipe, and golden plover were common along the grassy, and moist margin. Within the rushes, the sedate coot, with her black coat, and the curious white patch on the forehead, sailed with her little brood of seven or eight. The mallard was engaged with her second family. And the teal seemed to be playing at " housie," so diminutive, and matronly did she appear, and so burdened with the care of her twelve children. The tufted duck ; a few dun birds, with handsome red head and powdery grey back ; and a pair of shovellers, augmented by the progeny of the year, completed the apparent life of the lake.

Grouse came down from the hills to the roadside ; and, although it was already within sight of the Twelfth, some of them were so poorly fledged that I might easily have caught them with my hand. A stick would have been a deadly weapon. The prospects for shooting, therefore, must have been exceptionally favourable.

Along the fringe of the hills, blackcock were not uncommon, although the bird is on the decrease throughout the Highlands. Young sportsmen, of

whom there are so many, are charged with mistaking the female, or grey hen for grouse, and
killing her on the Twelfth, that is eight days before
blackcock shooting begins. But, as all the young
alike are grey, the destruction is probably much
more general.

Only a ridge separated the way along which I was
walking from Loch Tay and Taymouth, whither the
capercaillie, after a long exile, was reintroduced nearly
sixty years ago. A few had found their way across.

The carrion - crow abounded; the hooded crow
was absent. The raven was deeper among the
hills. Missel-thrush and mountain blackbird represented the thrushes.

The twite, or mountain linnet, was everywhere
on the hill slopes. It differs from the rest of the
linnet family in the absence of the prevailing red
tint. Ground builders are usually coloured in
harmony with whatever surface they habitually
rest on, and are destitute of any mark that would
call attention to them. There is a deep blood
patch on the rump of the male; but this only
serves as a further protection. It resembles a
flush of purple bells, among the purple heather;
this is the heather lintie.

Rain began to fall; so gently, at first, as only

to roll little dust pellets on the dry road; but
increasing till it dripped from the overweighted
leaves in a secondary shower of large drops; and
the lake became indistinct through the thickening
fall. This would never do; so I left what was
no longer a protection, and walked doggedly on.
It was rain of the wetting sort, each drop of which
seemed to cling. Erelong it fell with a distinct
noise through the air.

Gentle speculation beguiled the way as to the
nature of the prospective shelter. I could write a
book on those shelters. It appeared at length in the
form of a half-ruinous cottage, made up of the usual
" but and ben." The reception was on the north side
of friendly for awhile ; it often is in these parts.

Snug in a box-bed, I dropped over to sleep, to
the monotonous lullaby of two distinct drips
in the room. At what unearthly hour I know not
—for it is light all night in these northern parts—I
was startled into consciousness by a tremendous
thumping, accompanied by unnatural noises.
Bedlam let loose is the expression for it.

Her son was a wild sleeper, " but maype she
wouldn't be frightened, whateffer," explained the
old woman in the morning.

For the next four days I was in the position

occasionally depicted in *Punch*, and more serious
periodicals, under the heading, " A Week in the
Highlands." It rained well - nigh incessantly.
But a man is not, or ought not to be, the
weather's slave; and I was as much abroad as
the circumstances allowed, drying my single suit,
in the intervals, at the kitchen fire. As the only
concession to usage, I kept a certain shelter over
my head, and went to sleep to the sound of
the same two drops. After the first night, the
battle royal on the other side of the wooden
partition no longer disturbed me ; so readily do
we adapt ourselves to circumstances.

The loch stretched out in front of the small square
window, and was separated from the cottage only by
a narrow scrub of alders. It lay in the course of a
mountain burn, coming down a wild glen; filled the
usual stony basin, surrounded by the weeping hills;
and gave rise to a very considerable tail-stream.

In common with all the other Perthshire waters,
except the few in the extreme south-west corner,
which form the romantic Forth system, this one
found its way into the river Tay ; and, like all the
other Perthshire Highland lochs, except the same
inconsiderable Forth system, this formed one of
the group around Loch Tay.

Seven-eighths of the lochs of Perthshire cluster round this central brilliant, and seven-eighths of the streams serve to bind these into one, leaving only about one-eighth for the infant Forth. To have the centre of all the romantic beauty and interest summed up in the very words Scottish Highlands—for such is Perthshire—before him as a picture, one has only to be familiar with a few such labour-saving facts as these.

Fishers, like snails, come out in force during, and after rain. It would almost seem as if they forecast the change, so rapidly do they appear in the remotest country districts; dotting every stream-side, where yesterday there were none of them; and dipping their worm into every eddy and slackening of the tumbling water. There they were, wading through the wet grass, hopping over the dry stone dykes, and, generally, deporting themselves with an enthusiasm worthy of a better cause.

I inspected several of the baskets of those who were returning from the loch, and was struck, by no means for the first time, with the number, and insignificance of the trout. One good fish would have been more satisfactory than the whole lot. Fifty of those I saw might have weighed 10 or 12 lbs. Divided by twelve, they would have yielded an average of a pound apiece, and made a noble basket. Everyone who has tried, knows that the supreme moments in a day's fishing are those which intervene between the hooking, and the landing; while the latter is by no means a foregone conclusion. But, in this case, very little time would elapse, very little excitement could prevail, and no true sport would be enjoyed.

Such a noble sheet of water ought surely to have done better than that; and the weakness is by no means confined to it, but is common to far too many Highland lochs. The cause must be one of two: either over-population, or scarcity of food. The fish, such as they were, seemed well enough fed. The want was of some agency to keep down the number.

Nature's agent in the loch is the pike, just as her agent on the hills is the eagle; and to kill out the one, is just about as wise, and as much in the interests of true sport, as to kill out the other. There may be some lifting up of the hands at this; nevertheless, it is a view which many votaries of the rod, besides myself, would fervently commend to those who over-sedulously net their waters; or who, in any other way, seek to improve on the ways of nature, and establish a balance of their own. All such are agreed that they would rather fish where there are pike, and take what they have the skill to get, if it is only one good trout, than trifle with small fry. One cannot help wondering what sort of world we would have had, if some people had been at the making of it; and what sort of waters, had these same wiseacres been consulted about the stocking of them. An illustration occurs to me.

Some years ago I had the questionable privilege of fishing an artificial lake of considerable size, in one of the Highland counties; and found the yield to be not only disappointing, but even saddening. Because of their numbers, the trout were small, and from lack of sufficient food to support so many—for lakes with an adequate larder are not made in a day—they were thin as well. The best friend of the proprietor would have recommended some controlling force.

Pike are extremely catholic in their tastes; and are not given to discriminate between the young of trout and their own young, but feast indifferently on both. Equally catholic are the trout; who are by no means such helpless victims as they are represented. A two-pound trout probably owes half his weight to young pike, and thus succeeds in balancing the obligation, and getting good out of evil.

Once upon a time, Loch Katrine was over-stocked with small trout. Pike were introduced, with the usual result of larger, if fewer fish. This state of things continued for awhile, to the unbounded delight of the native angler, who could depend on a fish worth playing, and taking home. But it was not equally satisfactory to the less skilful tourist, who complained to the hotel-keeper of the poverty of the water. And, some scheme is now being anxiously waited for, which will abate the so-called nuisance. Happily, pike, once introduced, are not so easily got rid of. And, on such a large expanse, sufficient will remain to prevent a return to the old state of things.

A similar lamentation in the interest of numbers *versus* weight, is raised over the condition of Loch Awe, since pike found their way in.

The Loch Awe trout is said to be able to keep its own loch in order; and no lake which possesses this trout has any need of pike. It is described as neither so large, nor so shy, as the lake-trout; though it may reach 16, or 18 lbs., a respectable weight even for a ferox. It is familiar to the Highlander as the ghiroo, probably a shortening of gillaroo. But, does this distinction between ferox and gillaroo hold?

Some trout were forwarded to me from a small tarn or loch on the shoulder of Benmore, near Inchnadamph, in Sutherlandshire; whose stomachs presented somewhat of the toughness of a gizzard. These are said to be allied to the gillaroo of the Irish loughs. The characteristic thickening, which takes place in the middle coat of the stomach, is the very natural result of being shut up in lakes where the food is chiefly molluscs, whose hard shells have to be crushed up. It is likely to occur to whatever fish happen to be in the lake, and does not mark off any distinct species. If small, they are, as in the case of these Sutherland specimens, common trout with thickened stomachs. The large ghiroo, found in the Scots lochs, and known as Loch Awe trout, is probably the Salmo ferox modified by its food.

The rain did not prevent me visiting the Braes of Balquhidder,—Rob Roy's country,—and lingering round the edges of Loch Voil; nor from passing down to the singularly long and thin Lubnaig, more like the overflow of a stream than a loch, running along one side of Strathyre. These two lochs are interesting, as occupying an intermediate position on a sort of ridge, between Loch Earn, and the south-western system. Only, after a little seeming hesitation do they at length decide to pour their tail-stream into the Teith, thence into the Forth. Early spring fishing in these lakes is not good.

Dropping down to the Teith, Loch Vennachar is early; Loch Katrine uniformly good throughout. On the infant Forth, Loch Ard is early; Lochs Chon and Dhu are late. "Those inhabited by pike," says a veteran, "are my particular favourites, especially when the greater part of the shore is so clear of weeds as to make one independent of a boat."

The hills had the usual rounded summits, smoothed no doubt by the same grating glaciers; which, sliding down the glen, marked by the hill burn, ploughed out the basin of the loch. But there was one singular exception. A climb of a few yards behind the cottage brought into view a sharply-pointed conical peak. It was already the fourth morning ere

> The sun his beacon red
> Had kindled on Ben Voirlich's head.

This was the signal to depart. My further course led me amid haunts and memories of the deer.

XI

THE STAG

DEER are our big game; the interesting relics, retained through a process of preservation, of larger animals than any we now possess.

The roe most nearly approaches the wild condition. His small size, and ungregarious habits enable him to lead a comparatively free, and roving life. He is spread over the Highlands, and thinly scattered through the Lowlands. The fallow-deer, albeit degraded into an ornamental animal, still wanders in some of the ruder parts of the country, beyond the park fence, and seems as well able to brave the winter, and look after himself, as the roe. Both demand the nearness of cover for shelter.

The stag is the natural head of the family; and holds, undisputed, the proud position of premier wild creature—wild, I am afraid, by courtesy—of Scotland.

In addition to his personal picturesqueness, and to the suggestions of grace, strength, and speed, he has a certain literary, and national interest, which must wed him for ever to the soil. The new era in the life of the Highlands opened with a deer chase. And, all that has changed, and brightened the land since, dates from that opening.

The hue and cry arose in the neighbourhood of Comrie, swept down Glen Artney to Uam Var, turned westward over the site of modern Callander, echoed along the valley where lie the three mighty lakes, and only died away in the *cul-de-sac* of the far-famed Trossachs. It seems ungrateful, therefore, to talk of banishing the four-footed benefactor, which still lends interest and animation to the scene, and whose shadowy footsteps entranced thousands now trace.

The hill-sides were not yet deforested to any considerable extent; and the howl of the wolves was only too familiar.

"In the time of James v.," the Fitz-James of the *Lady of the Lake*, "their number and ravages were formidable. At that time great tracts of Perth were covered with forests of pine, birch, and oak. All these clouds of forests were more or less frequented by wolves."

Under these circumstances, the stag would not be likely to abate any of his natural watchfulness, or lose aught of his elasticity, wind, and resource. He would pass the night in that light, and wakeful sleep common to creatures accustomed to a rude awakening; and shake off the dew-drops in the morning, with the deliberation of one, too experienced in danger to be flurried by the scents and sounds coming down the wind.

There is no reason, that I could ever discover, why all the keenness of the chase should be confined to the hunter, and the dogs. More probably, it is shared in an equal degree by the quarry, when it is really wild, and not confined, and cowed, and tamed till all the ancient spirit has gone out of it. The fierce gratification of gaining on the pursued is only equalled by that of gaining on the pursuer. This consideration humanizes all real sport. If so the deer that day had the best of it.

> 'Twere long to tell what steeds gave o'er,
> As swept the chase through Cambus-More;
> What reins were tightened in despair,
> When rose Ben Ledi's height in air;
> Who flagged upon Bochastle's heath,
> Who shunned to stem the flooded Teith.
> Few were the stragglers, following far,
> That reached the Lake of Vennachar;
> And, when the Brig of Turk was won,
> The headmost horseman rode alone.

Such were the bold features of the day, when the stag was so little hampered in his movements, or bound to one place, that he drank one night at a rill in Glen Artney, and cooled his heated flanks the next in the waters of Loch Katrine.

Sport such as this is out of date. The horn is hung up in the hall, amid the other relics of a bygone time.

In any remnant that may remain in modern coursing, the better half of the chase has been thrown away, and only the baser part retained. What we witness is mainly a contest between the deer and the hound, in which the hunter is a passive, or rather excited looker-on. And, as the stag is by no means so spirited an animal as its progenitors were, nor so strong and speedy, nor in any way so well fitted for the ordeal, it is rather a poor affair at the best. A trial on unequal terms is not an elevating spectacle.

I cannot help feeling that there is a certain barbarism, when man is not heated by personal participation, I had almost said personal risk, which degrades all such duels between animals beneath the highest level of sport, and places them down beside similar contests in the arena. Somehow, they do not seem to fall in with our insular notions. They

are calculated to awaken in the sportsman a little
of the bloodthirstiness of the dog, without the
dog's keen relish for the chase. For the time,
perhaps, the dog is the nobler creature of the
two.

Stalking remained, whose strength lies less in
outrunning, than outwitting the prey. This
practice is adapted for scenes where the herd, or
the careless sportsman can be commanded from a
distance. It is so obviously the natural resource
of men with rude weapons, in dealing with such
a wary, suspicious, and fleet creature as the deer,
that it is probably of primitive origin. It preserves
many of the robuster elements of sport, taxing
the physical endurance, and calling out the
mental resources of accurate observation, and rapid
reasoning.

The object is to arrive within gunshot without
being seen, or suspected. To accomplish this, it is
needful to be familiar alike with the district, and
the game. Deer, grazing on the slope of a hill,
look down. The approach, therefore, must be made
from above. Even so: it is desirable to keep as
much as possible of the body out of sight, since
it is only reasonable to suppose that they must
sometimes look up.

The mode of progression down a steep is to lie on the back, and go feet foremost. The chief instruments of propulsion are the elbows; the heels must be used with discretion, lest the knees be raised into view. Meantime the rifle must be managed somehow. The distance to be traversed in this way may be two thousand feet, not all over soft ground.

Caenlochan, in the north-west of Forfarshire, surrounded by some of the highest mountains in Scotland, is as wild a little forest as any lover of sublimity can desire. It is a favourite haunt of mine for many rare things, besides its own impressiveness. Only here and there is an approach from the ridge safe. Innocent of any thoughts either of bloodshed or of danger, I attempted the descent. Commencing gently, and with an effort to walk upright, I paused at intervals to note the heightening skyline above, and the increasing savagery of the featureless scene beneath. By the aid of a glass, I watched the deer sheltering from the midday sun under the scanty shade, and could make out, from their motions, that they had winded, although, perhaps, they did not see me.

When too late to return, I discovered that I had mistaken the marks. The slope steepened so rapidly, that, in sheer self-defence, I took to my back, and half slipped, half propelled my body, I knew not whither. When the pace threatened to get too rapid, I steadied myself for a moment, by placing my foot against some projecting piece of rock, or by grasping at some mountain willow or tuft of holly fern. Though all's well that ends well, it was an experience.to pass through only once.

If this painful labour is not to be so much loss, the wind must be looked to; a comparatively simple matter in the open country, where, if it blows west at all, it blows west everywhere alike; but by no means so easy among the hills, where it has an awkward habit of changing its direction quite suddenly.

When on a level with the deer, the mode of progression may be changed from the back. The bare rough channel of the mountain torrent, or the shelter of the scattered boulders, together with every knot on the ground which will hide a prostrate figure, must be taken advantage of. Stony tracks must be crawled over on all-fours, with the action of a toad; or wriggled over, in a prone attitude, after the manner of a snake; while toil, and friction must be met with the stoicism of a North American Indian.

And all this is compatible with an ardent love for the surroundings. Without this, animal life, and all wild sport lose half their attraction. Indeed, there is little left now except the background. Whose mind does not revert to the pictures of Landseer?

"Deer-stalking with the Sutherland Highlander," says St. John, "seems an almost invincible passion. His constant thoughts and dreams are about the mountain corrie, and the stag. He points out to you with admiration the very mountain slope, the very corries, that you had already marked down in your mind as surpassingly grand. At first you may think him a reserved, and rather morose man; but when he finds that you are not only a brother of the craft, but also a fervent admirer of the natural beauties of his favourite lochs, and corries, his expression of face alters; he takes you under his protection, and leads you to points of view which you would have travelled fifty miles to see. Mercenary and greedy as, I am sorry to say, Highlanders in many parts of the country have become, I did not find this the case in Sutherland."

This is to be a sportsman, and this is what we look for, and look for in vain, in the modern devotees of the art. What we get instead is an absence of enthusiasm for the wilds, an absence of delight in difficulties overcome, which is the true charm and secret of healthy sport; a certain greed and vulgar boastfulness of mere numbers, and a love of flattery; the influence of which upon the ghillies has been to change the natural good-feeling, the air of high-bred civility, of which most mountaineers have a far greater share than men of the same rank of life brought up in the Lowlands, into servility, rudeness, and mercenariness.

The labour, jolting, and frequent abrasions of the stalk, together with the need to exercise one's own powers of observation and resource, are avoided; while very much larger results are obtained by the modern drive, in which the sportsman is skilfully disposed, with due regard to the wind, and every other contingency, and told to await developments.

All the available idle men in the neighbourhood are taken into the service, whose duty it is, by hallooing, and every other device known to them, to frighten the deer into a mass, and to force them on in the direction of the passes where the guns are posted.

Cautiously, as if suspecting danger, a few heads appear, and then the main body, pressed on from behind, follow. Thus the work begins. A true sportsman will select his stag, and try to confine himself to it; but, where so many are crowded into a narrow space, he is not unlikely to include a hind as well. And what of the indifferent shots, who, through excitement and want of skill, fire wildly? Useless slaughter is too often the result.

Many excellent men, whose devotion to sport cannot for a moment be questioned, have something to say in favour of this; but it is always of the nature of an apology, and lays itself open to the criticism, "Methinks the lady doth protest too much."

The deterioration of the game has kept pace with that of sport. Even in the early years of the present century, still more so further back, deer wandered at large. This led to a free mixing of the herds over considerable distances, and to that frequent crossing needful to vigorous vitality. Moreover, besides having free access to the sheep grazing, they descended from the snow-covered slopes, or emerged from the glens, to feed in the Lowlands, if, indeed, some of them did not remain there altogether. This prevented the periodical starvation, which may now be regarded as a feature of their winter life.

But, when the Highlands became the world's sporting-ground, it was discovered that they had a certain money value. Added to this, no doubt, complaints of their depredations were urged. And so the deer were herded into certain natural enclosures, where, from their stay-at-home habits, they settled down, equally to their own loss, and the loss of sport. In some forests, a modern system of fencing has been adopted, to keep them within bounds.

Thus severely isolated, and separated probably by many miles from the next herd, there began that in-and-in breeding, which led through the usual processes of degeneration. The deer became smaller in size, while tine after tine dwindled and vanished from the antlers. Only two or three of the past season's trophies will stand comparison with those even of twenty years ago.

The word "forest," as applied to their haunts among our mountains, does not bear its ordinary meaning, as it does elsewhere. It may be a survival from a former condition. The home of the Hungarian stags differs, very materially from that of the Scots deer. The more or less treeless forest of Scotland is replaced, in the first-named locality, by superb woods of deciduous, as well as coniferous trees; in the latter by dense pine, fir, and larchwood.

Amid all the differences in our forests, they agree

in the bareness of their aspect. Those with which
I am best acquainted are rough and untamable
glens, whose sides are scarred by the mountain
torrents, whose bottoms are uneven with tumbled
rocks; overgrown with the rough grasses, which
afford the coarse and not always nutritious grazing.

One advantage, if such it be, is to give the choice
of the season to the sportsman, not to the deer.
Where there is extensive cover, the stag must
declare its whereabouts before shooting is possible;
which it refuses to do until the rutting season
in October. Therefore stalking, or whatever other
form sport may assume in Scotland, occurs in the
pleasant days of August, instead of the uncertain
weather of the later months. Who would care to
come when the hills are mist-capped; when the dry
channels furrowing the slopes are streaked with
tumbling water; and the glen burns are in brown
spate?

Under the ragged patches of birch, fir, and larch,
the teeming herds gather, thus intensifying the
evils of overcrowding, and hastening on the deterioration. December hind-shooting has been adopted,
with a view of thinning out the herds, and so
mitigating the evil. And he who cares to join in
the slaughter, has a chance of seeing what the

pleasant autumn Highlands can look like in winter.
If he has taste, and hardihood, he may like the
ruder aspect and experience the better of the
two.

To many, the hound is known only from the
frontispiece to Scott's poems, or under the canopy
of his monument. In the main, he is a privileged
inmate of the kennel; more frequently a noble
ornament of the hearth, or a picturesque companion
of decayed gentility. For lack of use, he was so
fast dwindling in numbers, that the places where
he was known to be, could be reckoned on the
fingers of the hands; and, if he has been rescued
just in time, it was solely from a sentiment of
regret to see so noble a creature pass out of
existence.

While thus dwindling in numbers, he was no
less rapidly enfeebling in constitution. "Like other
greyhounds, these dogs do not continue fit for
service for more than six years. The violent pace,
and the strains they are liable to from the nature
of the ground they run over, and the strength of
the animal they pursue, all combine to make them
show symptoms of old age at an earlier time of
life than most other hunting dogs."

This overstraining, which tells so severely upon

themselves, is probably fatal to the continuance
of a vigorous race. And the in-and-in breeding,
rendered increasingly necessary with lessening
numbers, will accomplish the rest.

A well-known Forfarshire breeder was disappointed, time after time, by the pups either being
born dead, or dying a few minutes after birth.
Anxious to ascertain the cause, he had some of
them opened, when it was discovered that a weakness in the circulatory system made independent
life an impossibility. This, the anatomist, who
examined them, attributed to the extreme fineness
of the breeding, together probably with the inter-
relation between the parents.

The downward pace might be slackened, and the
constitution reinvigorated, by a good cross—say
with the mastiff, or the fox-hound. But a cross is
not the original. Forms of life are not preserved
among us on fancy conditions; and the deer-hound
will probably insist on departing the land, with the
cause which first called him into existence, and
found fitting exercise for his powers.

It is with sport as with a great many other
things. It is not criticism that can kill it out,
but internal degeneration and decay. So long as
it is manly; so long as it observes the ordinary

spirit of fair play between man and beast; so long as the creatures are really wild, and share the fun, and the sportsman is an enthusiast, and not simply a shot; so long as it is attractive just in proportion to the hardships endured; not all the puritanic talk in the world will have the slightest effect. Frederick William Robertson used to say that war, with all its horrors, was better than the spiritlessness and meanness which often accompany peace; and sport may be a more innocent analogue of war. If a millionaire will spend his thousands on a forest, so long as he is honestly bent on sport, better that than letting it rot in securities, and rubbing his hands on increasing dividends. A man is worth more than a million.

But waiting shooters, and driven deer, and firing that insults the silent majesty of the hills, and has no background in the spirit; and boastful reports; offend the better instincts, alienate the sympathy of the onlookers, and are disgusting and driving elsewhere all true sportsmen. If it is to be done in any way, the sooner it is undone the better.

It is not the crofter who will take possession of the forest, nor sheep that will banish the deer. It is the greed of proprietors; the omnipotence of the dollar, in whosesoever hand found; and the pity of mankind for "dumb driven cattle."

"When, finally, with palpitating heart, every fibre in your body set, you have approached your noble quarry, brace yourself by a supreme effort for a steady aim; a good stag is worthy of your very best effort.   And, if you are a true sportsman, and not merely a slayer, stay for a brief moment or two before you end that life, your finger pressing the trigger.   The call of a distant foe has just struck the ear of the gallant champion; and, with virile impetuosity, he steps forth from the circle of graceful hinds to hoarsely answer the challenge to mortal combat.   His head is thrust well forward, his shaggy neck distended to twice the natural size, his antlers of noble sweep are thrown well back, one of his fore feet is angrily pawing the ground, whilst his hot breath issues from his nostrils and open mouth upon the frosty air, like so much steam; it is a picture you will never forget." [1]

It is a picture which, unfortunately, one must cross the Channel to see.

[1] W. A. Baillie-Grohman, Badmington Library.

XII

AMONG THE BORDER STREAMS

THE southern uplands, if less popular than the north, are, probably, more beloved by those who frequent them. It was of this district that Scott made the remark, " If I did not see it once a year, I think I should die." The same spell took me back year after year.

There is no deer forest within its limits; perhaps no seclusion among the moderate heights absolute enough to furnish the necessary conditions. The roe runs wild. The capercaillie is unknown, although there is no reason why it should not appear in the patches of fir which are beginning once more to diversify the hills, and mark the renaissance of the old Ettrick forest.

Grouse abound, as one would expect, in the land of brown heath; but are probably not so numerous on a like space, as in the north. Blackcock are the distinctive game of the south—"the land of shaggy wood."

There is less rigid preservation over the greater portion; and little attempt, possibly little temptation, to let the moors at fabulous prices. I am seldom stopped; never insulted. One watcher covers the area of ten in less favoured districts.

I have known the local banker or housebuilder lease a hill or two in the neighbourhood of the village for an off-day amusement, and leave them in the charge of Providence, the best of game-keepers. The hand trained to the mallet or the pen is not always of the steadiest; and the shooting under such circumstances was not deadly. I have watched the birds after the sportsman had passed calling for the assistance of Nature's physician, the hawk.

And, just as the preservation is not so strict, the persecution is not so virulent. Vermin has never had all its criminal significance. The hills are not high enough, nor the precipices stupendous enough, for the nobler wild birds. I am not aware of the presence of the golden eagle, except as an occasional visitor. The peregrine builds no nearer than the Cheviots. But the minor members of the group, such as the merlin, or moorland hawk, abound.

Perhaps the largest resident bird of prey is the raven.

I never saw,   nor heard of the
osprey fishing   the Tweed, as
it does   the Tay;
but it   has been

                              known, at
                              rare intervals,
                              to fish Loch Skene,
                              and may well pay
                              many an unrecorded
visit to that lonely, and seldom disturbed place.

13

The Alpine hare lopes over the summit of hills, few of which rise above the heather zone. It was on Dr. John Brown's Minch Moor, a dark, weird, and shadowy hill behind the village of Traquair, that I first noticed its characteristic, leisurely fashion of moving on ahead of the intruder. Equally familiar are the mountain fox, and the stoat.

The otter is on the Tweed; but, it has always seemed to me that the artistic connection between this creature, and its surroundings, which make them part of one picture, is wanting. The stream lacks boulder-fretted stretches, sudden deep pools, rocky banks; and is altogether too placid. There is something in a background, for such an interesting and distinctive creature. The natural home of the otter is the south of Scotland; and the paradise of the otter - hunter, is the Dumfriesshire Nith. Hunting, over the greater part of the Tweed, would be out of keeping. The noise would jar on the peace of that mainly unechoing valley.

The case is different with the heron, whose motionless figure is seldom absent from the burns of the lonely side glens; and presents one of those startling pictures, with which Nature surprises the wanderer.

If there is any tendency to preserving, it is on

the water, rather than on the land.   One has less
liberty on the stream-bank than on the mountain-
side.   The gruffness of tone, and the strength of
expletive by which one is startled out of some
day-dream, into
which he has
been soothed by
the musical
summer ripples,
makes him rub
his eyes, and ask
himself if he is
in the High-
lands.   The
water bailiff is
the southern
double of the
northern ghillie;
and the natural   enemy
of the Borderer,   who can
never be made to see that he has no right to his
streams.

"Hae ye a line?"—meaning written permission
—roared one of the watchers to an angler, who stood
on the banks of one of the main tributaries, motionless as a heron, except for the swing of his arm.

"Do ye think I'd fish wantin' ane?" replied the phlegmatic Borderer, without turning his head. "Whare is't?"

"At the end o' my rod."

Rude, but characteristic. Some of those men are descended from the old rievers.

Fishing is the predominant pastime of the southern uplands, as shooting is of the Highlands. To possess a stretch of the Tweed is the Border equivalent of leasing a moor; to be invited to join the fortunate lessee is like spending a week in a shooting-box. One takes a rod to Peeblesshire, or Selkirkshire, as one takes a gun to Perthshire, or Aberdeenshire. Of course, he includes a gun in his Border luggage, just as he puts a fishing-rod into his Highland luggage; but mainly for off-days.

The Border district is a land of streams, which pass, by a slow increase of volume, the contribution of a thousand tributaries, from moorland burn to salmon river. The exception is the Yarrow, which, at the base of the hills, forms the Loch of Lowes, and, a few yards beyond, expands into sad St. Mary's. At one time it may have had a connection with Loch Skene; in which case we would have the typical three.

The lake-fishing is thus gathered within a narrow

space. And yet, the place is so full of memories, and stories, that pilgrims come from a long distance, to see where certain jolly anglers exercised their art.

Loch Skene is at best a mountain tarn, yielding indifferently fed fish. Few care to climb two thousand feet, and pass through as ugly an array of peat hags as ever I saw ; except, perhaps, for the boast of having fished the highest lake in Scotland. Last time I was there, a single devotee was at work, amid a solitude, suggestive of glaciers that had passed over as if but yesterday. He was on his honeymoon, which he had elected to spend here, for the sake of this inestimable privilege. We called at the cottage in the glen, and found his young wife, half tearful, and half pouting, at his notion of enjoyment. He was eccentric.

Hard by are the lakes, compared with whose classic traditions, fishing on Loch Tay is raw, and a thing of yesterday. On a neck of land, between the two sheets of water, stood, and still stands, in an altered form, Tibbie Shiels'. I used to visit Tibbie in her last days, and hear her talk of Christopher North, the Ettrick Shepherd, and many another of her memory's guests.

Until recently, lodging was hard to get, either

here, or in the cottages around; and the solitude and charm of the place remained unbroken. But recently, a grand hotel has sprung up, which daily empties on the loch a type of visitor, quite new to the south. The proprietor is concerning his soul at present about the pike, and desirous of killing them out, as if they were so many beetles.

As a fishing river, the Tweed is ideal: a succession of streams, and slowly deepening pools; with no traps for unwary feet, or ugly black holes of unknown depth. Wading is so safe, and easy, that it has become far too general, and disturbs the water for those who follow.

Once I saw a man wading, as I thought, recklessly, and spoke to him. He was there for a holiday, to recruit after the worries of the year; and there is no place more restful. His boy was playing on the bank. The water was high, and brown. He would cast always one yard farther. Afterward he disappeared; doubtless, he stumbled, and the water poured into his fishing-boots like so much lead; or the enclosed air made his wading breeches a couple of bladders to keep his legs up, and his head down.

Such occurrences, however, are exceptional; and, with ordinary caution, would be impossible. And

the truth of the ballad, which John Ruskin, somewhere, selects as typical of all that is best in ballads,
still holds :

> Tweed said to Till,
> What gars ye rin so still?
> Said Till to Tweed,
> Though ye rin wi' speed
> And I rin slaw,
> Whare ye droun ae man,
> I droun twa.

The best trouting, as well as the most picturesque
stretch, is from Peebles to Abbotsford.   On the
way, every pool has a name, borrowed from something characteristic.   Each envious, or incredulous
listener knows what is meant, when one boasts of
having hooked a three-pound trout in the Nut-
wood pool; "but just as I was landing him, you
know "—

Under ordinary circumstances, August is not a
good month for day-fishing.   But in the evening,
partly because of the raining down on the surface
of the dancing insect life, the trout become lively.
Toward eight I start for a favourite stream and
pool; and, as I pass along, the investing mountains
become more shadow like.   Arrived on the scene,
I find that the smaller trout are jumping, some of
them clean out of the water; a sign that I may make
my preparations in a leisurely manner, only hurried

by the increasing difficulty of seeing what I am
about. The larger trout at length begin to rise
silently, and in a business-like fashion, and the
circles to break on the surface without visible
cause.

A quarter of an hour later, and I only imagine
that rings are there; and can no longer follow the
flies, as they depart on their mission. A splash,
a run on the line, a prolonged birl of the wheel; a
pause, another run, another splash, a little stubborn
resistance, a few turns of the reel; another, but
feebler rush, all enter into the minutes of excitement
intervening between the hooking and landing of a
fish; with the added mystery of the dark.

Some big trout are feeding under the trees. I have
a rise, and lose him; and, the sudden rebound of
the rod sends the line among the branches. I have
to sacrifice a fly; and, in the attempt to put another on—the main difficulty in night-fishing—
lift the line above the shadow of the hills. In
much the same way, the fish must see the hook
against the sky, with what remaining light may
linger there, long after the fisher loses sight of
it on the dark water.

Standing breast-high among the sedges, casting
over the dimly seen surface, and listening, half-un-

consciously, to the muffled voices, and swish of
lines, and stumbling of the wading fishers, at the
tail of the next stream ; and the startling night cry of
a pheasant from the woods ; I feel a large creature
brushing past, within a foot of me. The otter has
come out of the water to pass the waders; and is
on his way to stiller pools farther up. The dance
becomes merrier. How that fish splashes! It
needs a fine touch on the line to know that he is
out of the water, so as to get down the point of
the rod in time to prevent a catastrophe. At
length, the frequent rugging ceases. Only a few
big fish fall at intervals on the pool, as if from a
high leap. The rise goes off for awhile ; probably
to recommence later. But, I leave the rest to those
ardent fishers, who are out for the night, and will
go home just in time for breakfast, and work. It
is one o'clock when I climb the dyke, and take
my way back, amid the sevenfold shade of the
tree-covered road.

A flood, after a rainless month, yields a good
basket. It scours away the accumulation of
bottom food ; and probably also, stirs the fish out of
the sickly, half-lethargic state into which they
had fallen. Such is almost the only condition
under which day-fishing is productive at this

season. The trout escape out of the sweep, and gather into the eddies, often far from the ordinary channel of the river.

If the wind blows down from the source up at Tweedsmuir, the water often rises several feet before the clouds are overhead; and cases are known, where anglers have been surprised in some shallow in the centre of the stream, and exposed to considerable danger. All night it rains, and all next day, and by the second morning the Tweed has overflowed its banks. Rain ceases, and the water begins to fall, as rapidly as it rose.

I take my stand by a grassy hollow,—where lovers are wont to sit in the twilight, and repeat the old, old story,—now covered over with brown water. The life of the adjoining pool and stream seems all to be assembled here; and, to have forgotten the caution which makes their capture, under ordinary circumstances, a well-nigh impossible thing. No sooner is the bait in than the rug comes; and such a rug! The excited fish makes for the current, and is swept down, carrying the hook with him. A little stronger gut will not matter; and the next is held, and the next, and guided up the slope.

Trout retreat along with the shrinking stream. They are never taken by surprise, or stranded;

except in the wild glen burns, which rage down
without an eddy, and sometimes entirely change
their channel in the course of a single flood. There,
they are thrown out by sheer violence. I have
seen the rain-bearing winds blowing down one of
these glens, without touching the sources of the
main stream. The burn rose rapidly, and swept
along "acres braid," entering the unsuspecting
Tweed, flowing placidly along at summer level,
with the fierce and bristling aspect of a wild
boar.

The brown changes into wine colour, suitable
for the minnow, and then resumes its neutral
shade. But the increased volume, and scoured bed
are favourable for a few days' fly-fishing.

The first autumn flood is always an event, for,
with it, is expected the first run of sea fish.

The bull-trout rush up in such immense numbers
as to choke the smaller tributaries, where it is
sometimes almost possible to scoop them out with
the hand. There has even been a proposal to net
the Whiteadder to give salmon a chance of entering.
This is, probably, only the ordinary salmon trout,
with certain local peculiarities. Its grilse stage
is known as the phinoc or hireling. And I have
frequently hooked it in a younger stage, along

with the salmon parr, from which it is distinguished as the "orange fin."

Old hands, bent on larger game, are busy getting ready their stiff rods, which no arms seem fit to wield, except their own. Some of the gaudy fraternity, improperly named flies, resembling nothing in the heaven above, in the earth beneath, or in the waters under the earth, become familiar objects on the stream. There are various ways, known to the initiated, of angling for big fish; but, far be it from me to tell what I have seen.

When a salmon reaches the mouth of the Tweed at Berwick, he is usually in splendid condition. In April, and May he is well supplied with more or less digested herring; during the rest of the year with sand-eels, or other available food. Whether he eats any more till on his way down the stream again, has been gravely disputed. When caught and opened, during the interim, no trace of food is found, and the stomach is usually in a more or less puckered condition, presumably from want of use.

This conclusion involves that the fish which runs up in the winter, or early spring, and does not spawn till the back-end, remains in the river for eight, or nine months, absolutely without food. That he takes enough to supply his lessened need seems

beyond dispute, and a quick digestion accounts for
the rest.

Although the movements of the salmon, when at
sea, have not been very clearly followed, seeing
that he is seldom, or never caught, we know that,
while there, he leads a very active life. He is, by
no means, a coast loafer, but joins the other raiders
of the deep on their predatory excursions; and
when he returns, he comes from afar. Under these
circumstances, a generous diet is necessary.

From the moment he enters the river, however,
the conditions are altered. The immense waste is
reduced to zero. No more exertion is needed than
just sufficient to maintain his position in the runs.
It is practically a state of semi-active, or semi-
passive hibernation, comparable with that of land
animals, only not carried to the same extent; in
which he can afford to live largely on the stores
laid up in the sea, possibly against this very
time.

Salmon visit the rivers mainly for the purpose
of spawning. In the Tweed, the spring, and
summer fish seem to play for awhile in the tidal
waters, and then go out to sea again. From the
marks frequently seen, it is supposed that some
of them, at least, may have run in to escape the

seals. During the hotter months, no fish seem to go far up the river, or stay in the upper reaches, as in larger streams.

The scene, from any of the bridges, at the back-end, is lively, and essentially dramatic. Salmon differ from marine food-fishes, in that they pair, and deposit sunken eggs. Lying on her side, the busy female constructs her redd by fanning up the gravel with her tail; while the male hovers near, ready to give battle to any intruder. Jealous rivals, and hungry trout, form an interested circle of spectators.

Many find their way into side burns, with scarce sufficient water to cover their backs. The ease with which they are reached, under such conditions, illustrates the wisdom of the petition, "Lead us not into temptation." Only a sensitive conscience need rob a Tweedside man of fish diet; and such a conscience, in the matter of taking a salmon, a good many Borderers do not seem to possess.

Various poaching devices are practised, from the leister upward, according to the state of the water, and the action, and position of the fish. When the river is in spate, a trap is prepared for the running salmon, which could scarcely be excelled for simplicity, and ingeniousness. Two nets, one with a

small, the other with a large mesh—larger than
the diameter of the fish—are run out from the
bank. The salmon rush against the smaller
meshed net, loosely floating below the other,
carry it with them through one of the larger
meshes, and are caught in a bag.

Hatcheries have come to the aid of Nature;
but, whether Nature is sufficiently grateful, it were
hard to say. For unexplained reasons, some years
are more barren; but, taking one with another, the
supply is fairly well maintained. And certainly,
the weakness of some streams, notably the Tweed,
is not the paucity of salmon.

After spawning, the exhausted fish drop back
into stiller water, where they can remain motionless; the females to the deeper pools near the
mouth of the river, the males to the first pool they
come to. There they recruit against the flood,
often very long in coming, which will carry them
to the sea.

Meantime, the stream is low, and not over-pure.
Moreover, it swarms with the spores of the little
fungus, known as *Saprolegnia ferax*. This pest,
because of its economic importance, has been made
matter of special inquiry, and its life-history has
been worked out in this its chief stronghold.

These spores attach themselves to the salmon; send processes down into the skin, and out in all directions, until, to the eye, they present a very perceptible patch, which, under the lens, resolves itself into a dense network, or matting. From this mass arise certain upright filaments, which thicken out at the end into a club-like swelling, where spores are produced, and ripened, and shed for the benefit of other kelts. A much larger proportion of males than females are attacked, especially, in the early part of the season.

No case is known of a new run salmon bringing this disease into the river with him, nor of any marine fish suffering from it. The fungus seems to belong to the fresh water; and, the obvious cure is to get the kelt into salt water as soon as possible. But, that is much easier said than done. The afflicted fish is too weak, and spiritless to attempt the journey; even if the frequent low state of the stream were not a sufficient obstacle. At one of the hatcheries, bay salt is used, but a good deal would be needed for a river, not to speak of the objections that might be raised by other fishes. The matter will have to be left in the hands of Nature. Such comparatively narrow, and over-crowded streams as the Tweed, are not unlikely to remain permanent sufferers; although, for some reason or other, the Tweed is the worst of the lot.

Trout, though they by no means enjoy immunity from attack, are less susceptible; with the exception of Loch Leven trout, which is a pet species.

The Tweed, which takes its rise far from the Border, and flows through several Scots counties, is not even under the Scots Fishery Board, and is classed as an English river. The Yarrow, and other tributaries, are thrown in with the parent stream. Where are our perfervid patriots, who fire up when Scotland is called England? Here is a grievance of the first rank.

14

XIII

GROUSE AND PARTRIDGE

HOW delightful it is to climb an autumn mountain; to breathe one's self on the gentle incline of the pasture field; to vault over the dry-stone dyke, as if gravitation had lost its hold; to push through the fir-wood, and startle the black-cock basking in the bare places among the bracken; to continue the difficult, yet unexhausting ascent through the heather; and, finally, to emerge clear above all, in the presence of the undisturbed ptarmigan, who fears no other enemy than the eagle.

Such is the vision which cheers the weary city man through intervening days of worry, and work. No more continental tours, from which one returns as unrefreshed as he started; nor yachting excursions, where one lolls all day in much the same attitude, and stares at much the same sea; nor

even gymnastic feats on the Alps.   But colour, and
life, and exercise; fresh mornings; lunch on the
heather, two thousand feet in the air, with a
horizon of many miles;   and pleasant
evenings, recounting the deeds of the
day.   Why, the opening up of the
Highlands was like the dis-

covery of
new heavens,
and a new earth!   It
added immensely to the gaiety, the health, and
the simplicity of nations.

The proprietor was wont to shoot over this scene,
together with a few privileged friends.   This was

in simpler days than ours. His family had spent the winter in Edinburgh; a thing that never happens now. Or, it may be that he was member for the county; for the "carpet-bagger" was still in the future. But Parliament rose about the end of July; and it would have been rank heresy, bringing on the leaders the disaffection, and desertion of their followers, to have sat over the Twelfth; or even so near as to make the journey hurried, and inconvenient.

The stations were not the present scenes of bustle; the trains travelled more leisurely. There were no sleeping compartments; and, as he shook himself up, put aside his rug, and opened the window to let in the fresh morning air, the shearers were already abroad in the barley-fields of an early harvest.

If young, and eager, he was up bright and soon, and on the moor before the grouse had filled their crops with the tender tips of the heather; if an old hand, he, probably, took another three hours in bed. "It will generally be found," says one who preserved the best traditions of the sport, "that if two equal shots, upon equal moors, uncouple their dogs, one at five o'clock, and the other at eight, and compare notes at two in the afternoon,

the lazier man will have the heavier game-bag,
to say nothing of his competitor's disadvantage
from having fruitlessly wasted his own strength,
and that of his dogs, when many of the packs
would not allow him to come within reach."

As he passed through the fir-wood, his attention
was attracted by a large bird sitting on a branch.
This was the first intimation that the capercaillie
had spread his length. His gratification, at this
addition to his game resources, was not altogether
unmixed. It might lessen the number of blackcock; and, from a commercial point of view, the
new - comer was not fit to eat in anything but
soup.

His foot was once more on his native heath; and
the spell of the hills, and the freshness of the day,
and the glow of purple, together with the mystery
of hidden life about to be revealed, were upon
him.

He shot after dogs, and was followed by a
ghillie; probably, an old retainer of the family,
born in some cunning cup of the hills, of a race of
gamekeepers, conservative of all ancient practices,
and duly indignant at modern innovations. The
little company of men and dogs, blended with,
and were lost against the gigantic slope; the puff

of smoke, and crack of the gun, in nowise disturbed the magnificent silence and solitude.

Alone on the hills, he trusted to his knowledge of the scene, and the game for a bag; and felt the stimulus and delight of being thus thrown on his skill, and experience. He knew how to humour the birds, and how to modify his plans according to the weather. He did not take a straight line across the moor, as I have seen some do, for all the world like a man ploughing a field, shooting what rose by the way. He began on the out-skirts of the moor; contented with a light bag during the earlier hours, so long as he raised the packs, and concentrated them on some undisturbed portion in the centre.

He followed the dictates of common sense, even to the choice of clothes, whose colour blended with the shades of the hills. One found that a drab-coloured cap took him five or six yards nearer his

game than the lowest crowned hat he could pro-
cure. The shade helped to conceal him. Of course,
he exposed himself as little as possible, advancing
in a bent attitude, and taking advantage of every
inequality that would afford him temporary con-
cealment. He approached from below, not, as in
the case of stalking, from above; seeing that while
deer look down, and can be least readily surprised
in that direction, birds look up. Such devices
added zest to the enjoyment of one who was as
much a naturalist as a sportsman; and took fully
as much interest in knowing the birds as in shooting them. After three hours of incessant tramping
up and down hill, in which mayhap he had only
prepared the ground; he chose some convenient
spot on the sunny side, commanding a view of
the picturesque lake underneath, for lunch. The
meal was light; experience had taught him the
lighter the better. The least alcoholic of drinks,
the most digestible of solids; enough only to
refresh; and then to work again.

Unlike those who crowd all their pleasures into
one day, and tire of grouse-shooting in a week,
he came home for a few hours to cool and rest;
returning in the evening, when the sun had already
dipped down to the summit of the hills. And he

found sport on the moors, flushing purple in the slant rays, or shaded in twilight, as pleasant and remunerative as that of breathless noonday.

His bag might not be large, according to the modern standard; but his simple creed was, that a few brace, shot over dogs, were better than a cart-load slaughtered by other methods. And, as with an easy conscience, and unimpaired digestion, he sat down, pleasantly tired, at the close of the day, he found, as in the case of all true sport, that the retrospect was still pleasanter than the actual enjoyment.

The ways of the hills are no longer the same. The love of sport, for its own sake, is fast ceasing to be the motive power. Interest in the birds has only a quantitative significance.

The introduction of driving into Scotland met with some natural resistance from the natives. In many cases, the poverty, and not the will, may have consented. But the prejudice is not shared by stranger lessees; who are, after all, those who have to be reckoned with. The unsuitable nature of the ground in many of the smaller moors—wilder and more broken than in Yorkshire, the natural home of driving—makes shooting over dogs less a matter of choice than necessity. But, wherever

conditions are favourable, the newer are supplant-
ing the older methods.

The moors of Aberdeenshire, and Inverness-shire
are those principally devoted to driving in the
north. Several of the moors in the latter county
are large enough to accommodate parties of from
half a dozen to a dozen guns, and can be made to
yield as many as 500 brace in a day. The highest
record for one day in the present year, was 375
brace, shot on the moor of the Macintosh of Moy.
Ayrshire, in the west country, comes second in the
race, with bags of 250, and even 300 brace a day.

The methods by which those numbers are piled
up are totally deficient in picturesqueness, and even
in originality. It is the background which lends
a certain impressiveness, not naturally belonging
to them. The frightened grouse, roused from the
heather by a shouting crowd, seem, in the language
of a sympathiser, to grow mysteriously out of the
rock in front.

"The great secret of success in driving is to
select those places in the flight of birds where they
can best be killed. The author is in favour of
massing the guns, and making the birds fly as
concentrated as possible. All the 'butts' should be
in a dead straight line, should be close together, and

well concealed, being invariably placed within gun-shot of a ridge in front, when on a hill-side; and if on a flat, then the outside turf should be carefully placed at the front of the butt."

The birds may be driven backward and forward over a ravine, and the sportsmen, posted in the hollow below, get them both ways.

And this practice from behind a turf erection, where the birds come to you over a ridge, and within gunshot, before they are aware of your presence, is that which threatens to supersede the olden method. For this ugly change various defences are offered, the strangest of which may be termed the commercial one.

Grouse, it is said, have become a quite appreciable element in the country's food-supply; and it is desirable that the largest number should be hurried into the market when the prices are highest. But they can never be so important in this respect as salmon; and gentlemen do not kill salmon to supply the demand; they leave that to the tacksmen.

Or, the modern methods, by more effectually thinning out the grouse, tend to prevent the appearance, and spread of disease. And yet, the natural enemies of the grouse, whose presence would bring about the same result in an equally effective, and much more picturesque way, are ruthlessly destroyed.

Attempts are being made to give grouse a still wider distribution. Other nations have caught the fever, and, since it is not always convenient to come so far, are eager to have their autumn-shooting nearer home. It seems a pity, from the naturalist's point of view, that a type, so markedly the creation of our peculiar insular conditions, should be expatriated. But, it may be interesting to watch how the continental colonies depart, more and more, from the parent stock, until they leave us our British grouse to ourselves, as before.

Why a day so early as the twelfth of August should have been chosen, or should be persisted in, for the opening of grouse-shooting, it were difficult to explain; if it be not that the heather is in bloom, or that the weather is likely to be good, or that the birds will be easily knocked over. We have no parallel to this, except rook-shooting in May, while the young are still on the edge of the nest, or perched, insecurely, on the branches. But then, that is a juvenile pastime. Not till the first moulting in October do the young cocks declare themselves, and become fair game. If this is rather late for the big bags so eagerly sought after, and would only appeal to those who prefer a few perfect birds to many immature ones, then some intermediate date, say the first of September, is not too large a demand to make in the interests of sport.

It would then be necessary to leave an interval between grouse and partridge shooting, to enable the same sportsmen to be present at both. Might not this be an advantage? and would not the twentieth of September more fitly initiate the annual onslaught on Chaucer's fearfulle birde? The young are immature even then, the late broods specially so, needing little skill in the shooting, and yielding, one would imagine, still less satisfaction when shot. The ideal date, in the case of all sporting birds, is obviously the change of plumage, no matter how late it may be. But that is open to the objection of being hard upon the fringe of winter, when the clay is lifting at every step; and the unhardy would rather be in the snug home coverts among the pheasants.

Besides, the shooting depends upon the weather, in a secondary sense. It demands for a background bare fields, broken by the green of turnips. If the cutting is delayed for want of summer suns, or, from untimely autumn rains, its range is

immensely narrowed. And, on the first of
September, how many British harvests are still in
the future. One sometimes wonders how, in those
slow-going picturesque days of hand-shearing, the
sportsmen managed at all; but, then, they were not
so greedy.

Though the partridge season opens legally on
Monday of this year, 1895, still the weather has
been so unfavourable for ripening, and the crops
are still so backward, that there is little prospect of
shooting being general before the end of the month.
All that can be attempted is among the turnips,
always an unsatisfactory, and unremunerative
process, so long as˙ there is standing grain to
conceal the birds.

The partridge is the most familiar of our winged
game; at least it is best known to the greatest
number. Who has not been startled by the cry of
the cock, and the sudden rise of the covey from the
lee side of the stone dyke, or the edge of the field
pathway; and watched the heavy horizontal flight,
marked by periods of alternate beating and sailing.
It is associated in our minds with mornings which
have lost the enervating mildness of summer, and
have not yet taken on the chill of winter; but are
delightfully crisp and invigorating; with scenes

softly veiled in mist; with hedges strung all over
with gossamer, whose every thread glitters with
reflected light.   Perhaps, there is no picture in one's
whole mental gallery so delightful as that of which
the partridge forms the central object.

Happily, as yet there is no boom on this shooting,
as there is on that of grouse.   It does not belong
to the Highlands, and that is sufficient.   The
general public seldom hear about it.   It is not ob-
trusively made known.   The records of the "First"
in the daily newspapers do not appear in the same
large type, or occupy the same generous space as
those of the "Twelfth."   The sensational bags are
discussed only by the interested few; they do not
serve as advertisements.   There is no spirited
competition among those, who may have no other
qualification than money to tempt, it may be, a poor
landlord from the path of self-respect. . There is a
remaining trace of modesty about the whole thing.
It is still, in a degree, the sport of gentlemen.
Long may it remain so.   For one thing, the scene
is too near the house, where a foreign element
might disturb the family; and, being among fields,
it often involves the good understanding between
landlord and tenant.

On many estates, shooting is still over dogs.

Next to the partridge, Ponto is the chief figure;
and where driving is resorted to, to tickle the
vanity of a proprietor, or to please the younger
guests, who may be shots without being sportsmen,
it is some satisfaction to feel that the "butts" are
usually ordinary fences, and not turf erections,
suggestive of a rifle range; and the semi-circle of
louts, who herd the game on to the concealed
shooters, are genuine yokels, all savouring more or
less of the turnips.

The partridge thrives best where the soil is
richest, and where the heather is at a distance. If
the stronghold of grouse is Scotland, and the
north of England; that of partridge is England,
and the south of Scotland.

Along the base of the hills, and the edge of the
moors is a neutral zone, where the grouse grows
large, and shows the lighter field-pencillings of the
partridge; and where the partridge dwindles, and
darkens into the moorland hues of the grouse.

There is a still more intimate intermingling and
exchange of shades, induced by crossing. The
comparative unfrequency with which this has been
detected, is largely due to the very slight overlapping of their distinct areas, and the comparative
narrowness of the band they occupy in common.

Crosses among other members of the same group, whose modes of life bring them into closer contact; for instance the grouse of the heather, and the blackcock of the bracken; or the blackcock and capercaillie of the same fir-wood, are much more frequent.

Of the crosses which do occur, it is safe to assume that the large majority escape observation, and all are speedily obliterated by the ordinary natural checks. Of a supposed cross between the red grouse and the ptarmigan, whose areas overlap still less than those of grouse and partridge,—the ptarmigan being perhaps the most perfectly isolated of all our game birds,—Professor Newman says: "Information received from other quarters, induces me to believe that other examples have before now occurred."

It is conceivable that just as favourite shelter and diet tend to limit each variety to the place where these exist, so certain tendencies, such as the preference of the female for one of its own kind, prevent such frequent intercrossing as would tend to confusion. The relations of the game birds seem to be of this intermediate character; and the differences among them neither so marked as those which separate the plovers, nor so loose as those which yield the multiplying varieties of domestic poultry.

Partridge, grouse, and all the rest of them may be described as wild breeds of fancy fowl, in Nature's poultry-yard, which, without losing the power, have lost the desire to cross.

XIV

VERMIN

THE hill-side, the pine fringe, the pasture, and
the grain-field, together with the water-
courses which plough them, shelter other forms of
life than the grouse, the capercaillie, the blackcock,
and the partridge.

If there was a time when the wild life of the
land was prized for its beauty, or that it lent
variety, and interest to the scene, or because Nature
seemed to have tinted it into exquisite sympathy
with its surroundings, such æsthetic canons no
longer prevail. And, if there is one who considers a wood empty, when there are no longer any
wild creatures to cross ahead of him, or to come out
of their retreats, as he stands motionless against the
bole of a tree, his tastes have long since ceased to
be consulted.

Every Sunday morning I used to pass a sweep

on the road to the hills. I knew that beneath the
rude exterior there was an unsuspected beauty, a
very pure love of nature. Very enthusiastically
he used to talk about his wild pets; and when I
ventured to ask him if he ever brought any home—

"Na! na! They're better whare
they are. I just lie still, and watch
them."

I met him coming back in the
evening. He had been out the
whole day; and while the hill breezes blew the
soot away, the little drama passed before him,
yielding him all the simple delight he asked for.
He envied no man; and swept away cheerfully
another week in anticipation of another blessed
day.

The element of sport has introduced a new classification, and a new natural history creed, to which we are forced to subscribe, whether we will or no. The entire animal creation now falls into the three arbitrary divisions of game, vermin, and those intermediate forms, which rank as game, or vermin according to the nature of the ground, and the sporting habits, and traditions of the place. And the disposition is to treat vermin as if they were of no interest to the age, and had no further claim to existence.

The fox of the hill, and the fox of the plain are two well marked varieties. The latter is carefully preserved, and petted for sporting purposes; and coverts are carefully made or left for him. The former is an arrant poacher, meriting only short shrift.

> Though space and law the stag we lend,
> Ere hound we slip, or bow we bend,
> Whoever reck'd where, how, or when,
> The prowling fox was trapped and slain.

Still he is a picturesque nuisance, by far the finer creature of the two, both in body and spirit. If he is descended from the other; then, generations of life in the open have given length to his limbs, and freedom to his bound, and absolved him from the low, cunning, and skulking gait, begot of prowling round the covert, or the poultry-yard. But probably he is the more nearly allied to the original wild fox.

A gamekeeper among the southern uplands informs me, that, two or three years ago, he was in the habit of catching these hill foxes, and despatching them to the Midland counties for hunting purposes. A similar trade seems to have been carried on in the Highlands. In view of the many advantages to which I have referred, —greater size, longer legs, stronger and more muscular build,—it was naturally supposed that they would give a better run. But their superiority was confined to their own rough domain, while on level ground, the more diminutive lowland foxes proved faster, and longer winded.

The otter is the only other animal in the same intermediate position as the fox. On salmon reaches he is persecuted, under a somewhat exaggerated notion of the mischief he works. Elsewhere, he is encouraged in the interests of sport. If nowhere systematically preserved, it is because he has a genius for looking after himself. By day, he is out of sight in some cunning hiding-place, concealed by overhanging bank or root; and, if disturbed, can enter the stream under the surface of the water. As we speak of lowland fox, and hill fox, so we speak of stream otter, and sea otter; the latter growing bigger, probably because of more generous feeding, as he approaches the coast. The sea otter is common on the west coast. I do not suppose that the variety exists here.

No animals have suffered such long, and relentless persecution as the owls. This may be, partially, accounted for by the prejudice created by their nocturnal habits. Even the gamekeeper of the old school, who was usually a merciful man, shot them, and nailed the skin on the door of an out-house, beside the horse-shoe, as a warning to evil-doers. And, when remonstrated with, shook his head only the more stubbornly. Yet they are man's unpaid servants, and, with the doubtful exception of the tawny owl, do no harm to counterbalance the good.

Owls are Nature's mousers, hunting in the barns, round the steading, along the hedgerows, and across the meadow, everywhere keeping down what would soon become a pest. Introduced to the dwelling-house, the white owl supersedes the necessity of a cat; and a most picturesque cat he is. His silent flight, revealed only by the waft

of air, through the darkened lobby, has something
ghostly about it.

The short-eared owl goes farther from the
houses, and is the messenger to the field, and the
moor. The mice on Inch Connachan were kept
down by this species;   but when the place was
planted over, so as   to hamper the
flight   of the

bird, the pest increased to such an extent, that a
boat-load of cats had to be introduced, which
probably turned wild, and did more evil than
good. The value of the short-eared owl in this
case was equal to a boat-load of cats; each owl
being as good as a cat.

In the recent vole plague, when men were at
their wits' end, and suggestions of wholesale
poisoning were in the air, the short-eared owl

stepped in, increased, and, contrary to his wont, remained in undiminished numbers during the summer, amid the abundance of food. The not unreasonable assumption is, that if Nature had been left to herself, the scourge might not have happened.

There is a certain fatalism about the suggestion, that voles increase periodically in quite a natural way; and would vanish in their own good time, without the interference of the owls, which scarcely commends it to practical people.

If the short-eared owl is nearest in habit and appearance to the hawks; the kestrel is nearest to the owls. He is the day-mouser, and, notwithstanding that he carries on his beneficent work in sight of all men, his skin appears among the rest.

The larger birds of prey rank as vermin; from the harrier which quarters out the pasture-field for the partridge, and the osprey that hovers over the lake for the trout, to the golden eagle that poises above the ptarmigan, and the mountain top.

The hovering of the osprey somewhat reminds one of that of the tern; the descent is as swift, the aim as straight. The capture is still more wonderful, seeing that; whereas the sand-eels may be in shoals, large trout are solitary. He is the least harmful of birds—for fish are plentiful enough to yield him his modest share—and the most picturesque adjunct to a Highland lake. Not even the diving wild fowl are so characteristic. In reading the works of famous naturalists, it strikes one, disagreeably, that, while blaming every other robber, from the grey crow to the professional egg collector, they never lose an opportunity of taking the eggs, or the young.

A pair of common buzzards, such as I saw the other day sailing, in their characteristic manner, around one of the lower heights, and calling to each other across the diameter of a vast circle, adds a touch of wildness, if not also majesty to the hills themselves. And these common buzzards are becoming uncommon.

The case of the peregrine falcon, though not so bad, makes one long for the time when the customary building haunt of a pair was placed under the special care of the occupiers of the land, who were made responsible, by the terms of their lease, for the safe keeping of the noble birds, and their offspring.

Last, and greatest is the golden eagle. When he is seen to strike, in nine cases out of ten, it is the mountain hare; a species which, because of its great increase, has, on many shootings, become a perfect nuisance. If, afterward, he is found perched on a ledge against a background of stupendous precipice, it is the same mountain hare he has in his talons.

If he strikes a grouse, it is usually one that can be well spared. It were easy to show that, whereas the eagle slays his thousands, grouse disease slays its ten thousands; and the fell scourge, usually, begins with the weaklings. The number of wounded birds in the autumn, especially with so many unskilful shots at work, is enormous. Those who are in the habit of being abroad on the hills are familiar with the laggard, painfully struggling after the pack. I have followed one until it dropped; and picked it up, to find it skin and bone. Such are left to drag out a miserable existence, and contract, and spread disease. And such are the victims which the eagle, if he were allowed, would remove.

As yet the widowed eagle seems to be able to find a partner, probably from Norway, where his kindred still nest unmolested around the fiords; and that may help to account for the fact that, at Blair-Atholl and Caenlochan, where the Grampians are at their highest, he is still fairly common. But that has not prevented him disappearing from, or getting rarer in, many of his former haunts.

The crows are, I am afraid, all thieves, from the jackdaw to the raven; and most of them are murderers as well. This is an indictment whose strength can only be justified by its truth.

The brightest and liveliest of the lot, and indeed of all our native birds, are the jay, and the magpie. Both have disappeared from many of their former haunts, and are fast vanishing from · the land. Especially is this the case with the former.

The jay is a woodland bird. There is no more richly wooded part of Scotland than the neighbourhood of Birnam; and yet, the appearance of a pair, in a scene so altogether favourable, was thought worthy of a paragraph in a newspaper. Probably they would be shot.

The hooded crow, and his cousin the carrion have been well named the rats among birds. There is simply no diabolical act of which they are incapable. We shall meet them again elsewhere, but among the mountains they steal eggs, and kill young birds; and, certainly, richly merit being kept under.

The raven, an ominous, but a grand bird, though present in the wilder parts of the Highlands, is nowhere so numerous as the lesser crows, or so great a pest as these just mentioned.

The vermin among mammals are the wild cat, the pole-cat, the marten, and the badger. And very rank vermin they are esteemed to be; for whom, it seems, no plea can be urged, except, per-

haps, the foolish one that they add picturesqueness and variety to our living forms; and the presumption that, unless they served some good purpose, they would not be there. Besides destroying a few useful creatures, they probably kill and eat a good many undesirable ones, and help to keep even game from the evils of over-protection.

Concerning the wild cat, it is extremely difficult to get reliable information. All reports from unskilled observers, of which many have reached my ears, must be received with extreme caution. There is nothing commoner than for a tame cat to become feral. In two houses which I used to visit, the cats were in this condition, feeding mainly on various kinds of game caught in surrounding woods. Both were very large, as if thriving on their freedom, and fare. In this transition state, cats often breed in the woods, and raise a litter which have never known domesticity.

In the Highland cottages one sometimes sees a brindled grey, with thick-set tail, regular stripes, and generally forbidding appearance; and learns, on inquiry, that it is somewhat of a rover. This is probably the result of a cross with a genuine wildling.

Except, it may be, in the wilds of Sutherlandshire, where game are not so strictly preserved, or in the rocky north-west coast, the wild cat is extremely rare. In a county so rude as Argyllshire, it is practically unknown.

Two species of marten are reported, one with

white breast, the other with a bright orange breast, and known respectively as the beech, and pine martens. They are probably only different stages, or sexes of the same. The marten is the most agile of our wild creatures, passing over the ground with the sinuous motion characteristic of the family, and climbing with the utmost ease, and grace.

To the question as to whether he had ever seen a marten, a gamekeeper replied, "Often." This seemed hopeful, and, as he hailed from the wildest part of the Highlands, not unlikely. But more particular inquiry brought out that he meant the sand-martin. And there is reason to fear that will soon be the only marten Scotland boasts. The Duke of Argyll writes about the west coast, "All gone, but within my memory."

Whereas the marten has a pleasant aromatic scent, like the green pine-wood he frequents, the foumart (foul marten) emits, when excited, an offensive odour. Whereas the marten is sharp-sighted and agile, the foumart is near-sighted, and slow. Readers of Smiles' interesting biography will remember how Edward was assailed by a pole-cat, while spending the silent hours in an old ruin; and how the brute uttered cries which the naturalist interpreted as a summons to others in the neighbourhood. Many a night might now be spent in the same place without fear, or hope of such a visitor.

A golden eagle was observed to swoop down on the ground, and rise again, bearing some animal along with him. Suddenly, he fell a second time, but in a helpless fashion; and, when the observer approached, he found the eagle's claws fixed in a dead pole-cat, and the pole-cat's teeth fixed in the dead eagle's throat. This was some time ago. Now there is little danger even of the sharp-sighted eagle making such another mistake.

The last of the four is, on the whole, the mildest and most inoffensive, although he is credited with a partiality for eggs, and young birds. The badger owes his comparative commonness to his night habits, to his deep lair by day, and to his utter disappearance during the winter. He is present in all the Highland counties, and thinly scattered over the Lowlands. Still, he is on the way to extinction, if a little further off.

The laws on behalf of those forms which have no sporting significance, and are of no commercial value, need amending. The weakness of the present movement is that it largely confines itself to birds, which are never likely to want an argument in their favour, or an advocate to make use of it.

The same fair play should be extended to mammals also. The shield should be made large enough to cover our native tiger the wild cat, our native bear, as it is sometimes called, the badger, and those two grand weasels, the pole-cat, and the marten.

The somewhat indignant protest of an outraged love of wild creatures has not been altogether without its effect. A slight reaction has already set in, which is even affecting sportsmen, who are at the same time naturalists. There is good reason to believe that the destruction of eggs and wild birds, albeit of vermin, will be everywhere restricted, if not absolutely forbidden. The rest is sure to follow, if not then too late, and the old balance will be in some measure restored.

XV

AUTUMN BIRD LIFE

15th October.

I, OCCASIONALLY, hear nearly every resident bird break out into some suggestion of its spring song. A few seem to be nearer the singing point than others, but the rest only need a sufficiently strong incitement. A gush of warm sunshine, or even a spell of mild, though dull weather, will cause even the most silent, and unresponsive to forget that it is the silent season, and ignore the many snowstorms yet to be passed through before the days of song come back. Under ordinary conditions, however, I find the list of birds that sing, with any measure of regularity, between August and February, to be a very short one.

The accepted explanation of the bird's song is that it belongs to the breeding season, which may be roughly said to extend from February to July. "The male sings to charm and win his mate," says

16

Darwin. "The male is in a condition of excessive vitality at that time, and gives vent to his high spirits in song, just as happy, and healthy human beings are in the habit of doing," says Wallace.

The theories seem to be complementary, rather than contradictory ; and, on either supposition, we scarcely look for song in autumn, when the birds are probably moulting, and in indifferent condition ; or in winter, when they are under-fed, and often find it hard enough to exist at all.

The passionate strains of the lark may well be addressed to his patient partner among the bents. But his health must be of the most exuberant description to give him strength enough to throw away on so many visits to the clouds. It is hard to imagine a lark that has sat out all night in several degrees of frost, and then gone without his breakfast, besieging heaven's gate with jubilant melody. True, he was singing when I crossed the links yesterday, but only a mere fragment of his song, delivered from the ground, or at most from a few yards in the air.

Last December, I left the song-thrush silent in our Scottish wood copses, and found him in full song in the Isle of Wight. This would seem to admit of only the one explanation, that the starva-

tion rations, and the depressing atmospheric conditions of the North had given place to a warmer
climate, with comparative plenty; and the former
was the cause of the silence, the latter of the song.

The warblers cease singing during the remainder
of their stay with us, after the nesting; because they
are moulting for their long flight, and, consequently,
in indifferent health. But, when they reach the far
side of the Mediterranean, where another summer
awaits them, and a continued supply of their
favourite food, they, probably, resume their song;
although there is no reason to believe that they
mate, or nest afresh.

The chimney swallows warble on delightfully
till they leave in October. The reason for the exception seems to be that, unlike the other migrants,
they do not moult in the autumn; while the air
is mild, food plentiful, and all the other conditions
of song are present. Their mute season is in the
spring, which is their time for changing feather,
against the northern passage; when all the warblers
are vocal.

I know only two birds, in this latitude, which
seem able, amid adverse conditions, to maintain
sufficient health and spirit for winter song.

The robin is the autumn bird *par excellence*.

His colour blends so perfectly with the fading foliage, that, when he drops from the beech-tree to the ground, his red breast is indistinguishable from the falling leaves. His pure and delightful song, with its clear trills, comes from the roof, from the paling, from the hedge. As each one

seems to be vocal, it would be easy to take a rough census of the district.

He is an early riser. His voice is the first thing I am conscious of, after the chirpy scream of the awakening blackbird : the two autumn, and winter morning clocks that tell me the time. He is to the dull dawn of these later months, what the missel-thrush is to the spring. And, as he continues till the other begins again, he thus completes the cycle of the year.

Why the robin should sing when others are
silent, it were hard to say. I have tried to make
out for myself, and have inquired of others, but
without arriving at any very satisfactory explanation. He is a soft-billed
bird, like those
that leave us
before the

arrival of the
bad weather, and go
to sing elsewhere.
It may be that he
migrated once, and sang all year round as they do,
but got too lazy for the long journey, and stayed
behind, trusting to pick up a precarious living
here. Only he kept up the practice of song,

as an interesting survival of a previous condition.

Or shall we adopt the simpler explanation, that he is that familiar, not to say impudent bird, who, rather than be in want of anything, comes to the window-sill to beg, and so manages to maintain himself in tolerable condition?

I find my second singer, the water-ousel, down by the burn-side. It needs a practised ear to catch the strains amid the rush and tumble of running water, so exquisitely does it blend with the deeper sounds of inanimate nature. Like the current, the song is continuous, without natural beginning or end. Following the sound, the eye has little difficulty in picking out the white-breasted singer, sitting on a boulder, or under the bank; or the black-backed songster against snow, or white froth.

May he not be singing in imitation, or for sympathy? We have seen how his song harmonises with, and even mimics, the continuity of the current. Your cage-bird may be silent; but start the spinning-wheel, or sewing-machine, and it will go hard but he will outnoise you. Cease, and he will follow your example.

In some such spirit of rivalry, may not the bird, who is never out of hearing of the gurgle or ripple, whose flight is over the very centre of the dancing current, whose favourite perch is some stone round which the water noises and sparkles, sing all the year round? Other birds, which haunt the silent woods, or retired hedge-rows, may, in like manner, be dumb for sympathy, or want of stimulus. And, is not winter the very season when the bed of the stream is fullest, when the flow is swiftest, and the merry din is loudest?

Or, is it the never-failing larder of the water, whose current will not stay long enough to be frozen, whose surface covers the larvæ of all next year's dancing water insects, that maintains body and spirit alike at the singing pitch?

I do not say that these accounts of the robin's, and the water-ousel's song are to be taken without reservation. But that they do sing, while other birds are silent, is a somewhat singular fact, which calls for some manner of explanation. If one never asks himself questions, and tries to find answers, he will never discover anything; in addition to remaining an uninterested, and unintelligent spectator. A speculation, which has no other outcome than to incite others to find fault, and suggest a better, has still its uses.

And there are other problems besides these to

occupy vacant minds, to fill the leisure of busy
lives, and to rebuke all who are disposed to regard
the world as used up, and life not worth living.
Little things in themselves, no doubt, but wonder-
fully suggestive, and full of interest notwithstanding.   One can scarcely move between our
hedgerows, with their fading foliage, and scarlet
berries—noiseless except for the clear note from
amid the deep red of the mountain ash—without
having his senses filled with the surrounding
beauty, and his mind pleasantly stirred to curious
guesses, and led on to far-reaching thought.

1st November.

A late October gale stripped the trees in a night
—that is now nearly a fortnight ago. The farmer
is taking advantage of the open day to turn over
the soil in the field to the right, and in the wake
of the plough follow a multitude of birds.

Running along the fresh furrows at a great rate,
as if each were trying to outstrip his neighbour,
and secure the foremost place, which is probably
the case, and tumbling, and flapping over one
another in their hurry, are the starlings—hundreds
of them. If the peasant is an old man,—and he
need not be very old either,—he will probably tell

you that he remembers the day when he did not know that such creatures were in existence; and now their metallic hues are as familiar to him as the black livery of the rook, or the lighter shades of the gull.

Daintily picking his way along, with a run,—he is one of the running birds,—and a pause, is a single pied wagtail; thus showing that he does not quite desert us in winter. "The grey" will be found by anyone who is curious enough to cross the dyke to the stream beyond.

With a clatter of wings, and a headlong flight, not unlike the curlew's, some heavy birds issue from the strip of wood to the left, and light on the turnip-field on the opposite side of the road. The white patch on either side of the neck, though it meets neither above nor below, proclaims them ring-doves. The slaty blue back, and pink flush on the breast, complete the picture. These are they whose melancholy love notes breathe an air of sadness over the hushed summer landscapes: then they are scattered in pairs. In winter they gather into flocks, and in hard weather visit the turnip-fields to feed on the leaves.

A flock of chaffinches are busy among the straw of some open potato-pits. They are nearly, though not quite, all female, with perhaps a male in each score. Half a dozen males are feeding by themselves on the road, and evidently leading a sort of winter bachelor life. Thus, although there is by no means so absolute a separation of the sexes as to justify the specific name of " Cœlebs," there is the distinct tendency on the part of, probably, a majority of the males, to live, more or less, apart. I find that, as a rule, it is safe to assume some basis of fact in what old observers tell us.

Feeding along with the chaffinches are two blue tits. These bright little fellows press their tails on the ground, when they have a tough morsel to deal with, a habit doubtless acquired in their homes up in the fir-trees. In the odd position they have to adopt in order to reach their insect food, sometimes upside down under a twig, they doubtless find the tail a convenient aid to the bill. But they are not such masters of the art as the brown-backed creeper circling round that elm trunk, nor has their tail been modified into so apt an instrument.

A song-thrush pops over the wall, and lights on a fallow field. There is no mistaking him. He is too large for the redwing, too small for the missel-thrush, too brown or yellow for the fieldfare. Colonel Drummond Hay reports that this bird leaves the neighbourhood of Perth for the winter; and the Duke of Argyll makes the same observation about Inveraray. Plainly his migrations are only local.

The little dark hedge-warbler comes into view for a moment, and disappears, once more, among the confusing shadows of the twigs. A soft-billed bird, like the robin, left behind by the departing warblers, until perhaps he has lost wing power for so long a flight, even if he desired it, he lacks the boldness of the impudent little beggar of the window-sills. Be it snow or rain, he remains out in the open.

The winter form of bird life is flocking, just as the spring form is pairing; only the former is not quite so universal, since there is not the same overmastering necessity. Indeed, there are certain very obvious reasons, connected with the prime condition of life, why certain species should remain scattered.

The insect-eating birds which continue with us do not flock, nor even those that live largely on soft food. Neither of the insect-eating thrushes—blackbird and mavis—flocks. On the other hand, the berry-eating missel-thrush flocks. A not improbable explanation is that insect food in the larval stages is pretty much scattered; and, in winter, when flocking takes place, scant, and hard to get. A thousand hedge accentors, scraping among the withered leaves of one hedgerow, would find but starvation rations.

The seed-eating birds, on the other hand, which drop down on the stubble-fields, where the winds of last autumn, or the passage of the reaping machine have shed countless grains; or on the full stackyard, where the fruits of the earth have been gathered, and of which they take their share without so much as saying "By your leave"; or are contented with the plentiful supply of groundsel, or plantain, or grass seed which nature has spread over wild and waste places, congregate, because where one can find food, a thousand may.

The whole matter of winter flocking remains to be determined. In the case of migrants, which join in the passage, and scatter at either end, it has plainly some close connection with the dangers of the journey, and the safety which the individual finds in the mass. Some such motive may incline our seed birds, which are not kept apart by the stronger necessity of getting sufficient food, to live and move in societies.

The advantage of a winter walk is its delightful freedom from restraint. The fields are bare, and even the gateways have been left invitingly open, since the passage out of the last harvest cart. One is not bound to the high-road, but may turn aside anywhere, without fear of being rudely challenged.

A flock of larks rise from the stubble, and flutter away with their vibrating mode of flight. If this bird is decreasing, it is because of the cook, and the lessened area of breeding-ground. I do not think the cook counts for much in Scotland.

On the far side, the field passes into the links, with only a turf dyke to mark the transition. So long as we have barren stretches such as these, the larks are safe. Men do not plough the sand. The life of the bent is largely from home at present. A solitary stonechat poises for a moment on the extreme top of a furze bush, and then flits off to another and more distant perch. A solitary meadow - pipit rises from the grass, and drifts away.

The linties, or such a miserable remnant of last year's flocks as are left, were feeding on the fallow field as I passed. If the larks are safe because they build on the ground, which no one can remove, the case of the bush-building linties is different.

The olden charm of the links was the glow of

whins; and the redbreasts, and sweet twitter of pleasant inhabitants. Before the yellow crow-foot bloomed, or the blue sea-side butterflies were abroad, it was delightful, beyond expression, to look over the mass of colour, and listen to the merry sounds of the nesting season.

A craze,—and even a very legitimate thing, when carried to excess, becomes a craze,—spares nothing. The whins are burned, the linties are banished. We do not object to sport; we rather like the sunny side of life. A day on the bents helps to keep the spirit young. He who never relaxes is not only a dull man, but an uninteresting companion. But we like a little fair play between man and the inferior creatures; and, moreover, we think that the sunshine of life includes the glow of the bush, and the song of the bird, as well as the flight of the ball.

To the early players, golf was little apart from its surroundings. It meant getting away from the hard streets, and the smoky atmosphere of the town, to the springy turf, the fresh breezes, and the blue sea and sky. Among other things, it meant losing one's ball among the whins, and pausing for a moment in the search to listen to the lark, or lintie. It was a joyous game then.

30th November.

The winter equivalent of the warbler is the finch, including the bunting. It is he in the main who fills the vacant place, and gives a character to the months which intervene between the fall, and the return of the leaf.

A number of birds are feeding on the strip of links opposite St. Andrews Golf-house. Their motions are lively in comparison with those of native species; and, ever and anon, they squat on the ground, in a manner which no self-respecting Scots bird would think of imitating.

The flash of pure white in the expanded wing belongs to no resident, not even the summer chaffinch. Further proof of their identity is supplied as they flit a short distance in a sort of curve, and drop in a shower on the ground, to the sound of their sweet tinkling voices.

With a view of watching their motions more closely, I approach within a few yards,—a liberty no native bird would allow. And I recall that, in their abundance and tameness, these birds are to the Esquimaux what the house-sparrow is to us. Indeed, it seems to me that the strangers show a confidence in human nature exceeding even the impudence of the sparrow.

They find plenty to do in picking up the minute
seeds which the winds of last autumn have shaken
out of the grass pannicles, and leave me at liberty
to pursue my observations at leisure. The prevailing shade is tawny, relieved in the males by the
white wing-feathers, which entitle them, even in
their winter dress, to the name of " snow-flakes."

There are sixteen in all; a small flock compared
with that into which any of our resident species
gather. Three, although following the motions of
the rest, seem to keep together. These are about
the size and build of our common linnet. Their
characteristic is not white, but the dusky hue
round the bill, and the dark glowing crimson of
the crown and breast,—redpoles they manifestly
are,—the mealy redpoles of the North.

Each season, probably, the snow-bunting drops
a few behind to breed. "A pair was seen by
myself," says Colonel Drummond Hay, "evidently
nesting on the face of Ben Macdhui, above Loch
Avon, on the 21st June." The mealy redpole has
not yet been seen in summer, and, indeed, seems
only to be an intermittent winter visitor.

Another bright northern finch, the brambling
winters with us in varying numbers, and is suspected, although without very distinct proof, of

occasionally nesting. "They perch on trees ; and I have seen them more often about beech-trees than any other; but this may be the effect of chance." In appearance, they resemble the female of the snow-bunting, but differ in their perching habit.

Our own resident members of the same group scatter south, in numbers varying with the mildness, or severity of the season. All our finches, indeed, all our seed-eating birds which move at all, change from colder latitudes. All our native finches move from the Highlands to the Lowlands. Whereas, the soft-billed birds appear in the southern counties of England, and some refuse to go very far beyond; the hard - billed birds strike our northern and eastern coasts, or come from across the Tay.

The sky-line of the Grampians would seem to be the southern limit of several seed birds; for, although they may overflow into the Atholl Highlands, they are nowhere so numerous, or so much at home, as by Dee, or Spey side.

The siskin is one of those. That interesting little cage-bird,—which has nested, and raised its brood, or broods among the alders of mountain streams, or hill-sides,—comes out of his Highland fastnesses to the open grounds beyond, in search

17

of Lowland alders ; where he is found by his natural enemy, the bird-catcher.

His equally lively little companion, the lesser redpole, who has built, and spent the summer among the northern birches, appears in still larger clouds. I have seen the surface black, or rather brown touched with red, as the restless atoms fed on the ragworts which rose through the snow.

No other winter bird is trapped in greater numbers, or for less apparent reason ; except, perhaps, for his entertaining vivacity. Ingenious devices for compelling him to draw the water each time he wants to drink, are thought to add to his value. Everyone who has had a peep into the teeming cages of the out-at-elbows fraternity, must have felt that some restriction should be placed on the use of net and bird-lime. It is quite easy to clear out a flock of starving birds, and little care is taken to set even the useless hens at liberty.

A crackling among the snow-laden fir-trees, and a raining down of cone scales, tells of the rarer and more reluctant southern movement of crossbills.

Two interesting Scandinavian thrushes cross to the milder side of the North Sea, arriving here together early in October. The fieldfare is noteworthy as the only one among our thrushes that

builds in colonies, like the rook among crows. So
that its flocking may be said to be perpetual. The
redwing, because of his delightful song, is known
as the Norwegian nightingale.

If we include the two ousels, there are no less
than seven thrushes with us during the winter.
And, as no thrushes come to us from the South in
the spring, and none of our resident thrushes leave
on the approach of winter, we are justified in
regarding the type as a northern one.

The hooded crows have sensibly increased in
numbers. These birds are resident in the north
of Scotland, migrants in the south, and in an intermediate condition in the latitude of the Tay.

Some confusion has arisen between the hooded
and carrion crows. So identical are they in haunt
and habit, that, but for the grey of the one, and the
black of the other, it would be impossible to distinguish them.

"I have taken some trouble in examining these
birds," says Colquhoun, "and have killed crows
of every shade of plumage, from pure black to the
perfectly marked hooded crow; and this without
reference to age or sex. I have also seen a perfectly
hooded crow paired, and breeding with one quite
black." This may only mean that the species are

so closely allied, and so thrown together, that they frequently cross: a condition of things common, as we have seen, among game birds, and even swallows.

"Nests are constantly found with one of both kinds; and I have noticed that the male is generally hooded, and the female black. The young also are mixed," says St. John. This would seem to indicate that the grey, instead of being a specific, is a sexual badge, like the colours of other male birds: a suggestion which is contradicted by the fact that pairs of grey crows are found.

The hooded crow is a northern and coast species; the carrion crow, a southern species, whose range extends much farther inland. Where the two meet, which they do on the flat seashores, they undoubtedly mate.

If the distinction is to be observed, then they are the only members of the crow family that do migrate in this unmistakeable fashion; although the raven may come out of the stern corries in exceptionally severe weather.

Game are noted homekeepers. Indeed, they are the most truly stationary of all our native birds. The ptarmigan may descend to the edge of the snow cap; the red grouse may join the blackcock, and the blackcock may join the partridge in the stubble; the partridge may be driven from the upland fields to the richer, and more sheltered lowlands; and that is about the utmost range.

But, there are several important additions to the winter birds for the gun. At times every true sportsman must tire of conventional shooting. A moor is apt to get monotonous in proportion as it is well stocked, and a bag uninteresting in proportion as it is well filled. A hundred and fifty brace, all grouse, is often a hundred too many. The secret of true sport is not knowing what is coming next. The charm of emptying, as of filling a bag is the variety of feather.

One may find a freshness in the wild and way-ward game, which seem to be so much a part of the wild and windy days, with their freshening rains; and invade with gusto the spongy marshes, taking with the utmost good-humour the frequent stumblings over rush-concealed stones, and into hidden pools. The sudden sharp "scaip" as something rises, and, with a rapid flight, swiftly mingles with the grey; the quick shot at the disappearing shadow; the walk home in the fading light; the change from wet to dry garments; all make an ideal close to the year's inland sport.

Most important, from the sportsman's point of
view, of all those errant birds that are free as the
air, and wild as the wind, are the woodcock.
Scattered over the rough and broken ground, chiefly
in the moist patches indicated by the rank grass;
in open weather they shelter in the hollow furze
bush, or betake themselves to the birches, should
the rain change to snow. In Sutherland they have
commenced breeding in most of the woods, laying
four large eggs, similar in colour to those of the
snipe. The eggs of the woodcock are easily distinguished from those of the snipe, or any other
kindred bird, except, perhaps, the great plover, by
their exceptional roundness and fulness.

The visitors come in October. Their approach
is made known by the redwing, which bird one
cannot help connecting with the woodcock as
guests who commonly arrive together, however
unlike in other respects. A bag of woodcock, with
the richer winter plumage of the golden plover for
feather, and a mountain hare for fur, are pleasantly
suggestive of bracken, marsh, and coppice.

The flushed mire snipe tacks first to one side and
then to the other, and demands from the sportsman
a knowledge of his habits, to judge when to shoot.
After the third deviation, he settles down to his

steady flight. Though resident in abundance, many leave the country in spring. Probably, they, too, began by dropping a few behind, which number gradually increased.

The jack-snipe arrive in September, and it has never been satisfactorily shown that any remain. In time, they will probably follow the lead of the rest. There is a distinct tendency in our migratory birds to become resident.

XVI

WILD FOWL ON AN ESTUARY— WINTER

THE days are at their shortest. The plough-man's gulls, those seaside clocks, leave for the fields about eight in the morning, and return about half-past three. All day long the sands are left in possession of the black-headed gull, which is no longer black-headed, and can only be distinguished by his knowing look, his curious pattering run, and his red legs. If the day is frosty the common gulls do not leave; they seem to know that no ploughing is going on, and that the fields are as hard as iron.

Once every day life is at a standstill. With the full tide a silence comes over all, and a folding of the wings. The various forms, that made the sands so lively a while ago, have vanished as utterly as if they had gone out of existence; and are doubtless drowsily "peeping," and preening their feathers on some favourite rocklet rising out of the sea.

But, no sooner has the water ebbed sufficiently far to expose their feeding-ground, than the gulls take possession of the sandbanks; the redshanks announce their approach with a double scream, and recommence their restless search among the dripping seaweed for the clinging mollusc; the ring-plovers flit past in small bands at lightning speed; the dunlins drift along, wheeling in their characteristic fashion, and returning on their course, as if influenced by a sudden change of wind.

The sandpipers seem to mingle freely, so that it is very difficult to tell what species may, or may not be represented in a passing flock, without the test of the gun; and, when one has fired, he never knows what he may have killed until he picks it up: it may be dunlin, purple sandpiper, sanderling, godwit, or knot.

When big and little flock together, they seem naturally to fall into a certain order; godwit leading in a compact body, knot coming second in a looser flight, and the great army of smaller sandpipers behind. Hidden by numbers, no doubt, a great many species escape undetected.

The knot is one of the most interesting of our

coast birds. He is a winter migrant. The exact limits of his summer nesting-ground in the North remain undetermined. His wing power is wonderful. His range is probably greater than that of any known bird. He leaves the Arctic region for a winter home beyond the Equator, travelling, by the way, at great speed. He drops contingents as he goes. From November till March, he swarms along our coast, and then disappears utterly. He is one of our most graceful shore birds, with a short black bill, and about the same size as a godwit.

A pair of silver plovers light ahead. If their soft winter shades are so pleasing, what must their summer plumage be? They are called solitary, because each is supposed to prefer his own society; but they are, frequently, met with in flocks of from eight to a dozen.

A sound like the grating of a rusty wheel comes down from a V-shaped flight of pink-footed geese. These are usually night-feeders, keeping watch and ward by day against the approach of all suspicious characters. But in the absence of moon, they sometimes reverse this process, leaving the banks at sunrise, and returning at the dusk.

The coast takes a bend to form the estuary, across the mouth of which, the water breaks in a long silver line on the bar. The quavering whistle of the curlew, which has been audible for some time, and, under favourable conditions, should carry more than a mile, becomes increasingly distinct. The lower sleepy winter "peep, peep" of the oyster-catcher, which one must be near in order to catch, is now added. These sounds are never absent, except when the tide is full, and the daily hush comes on. Then the curlew is feeding in the fields, and the oyster-catcher resting on the moor.

A few moments and the birds themselves come into view.

The oyster-catcher patters along the very edge of the water, "peeping" drowsily the while; and in flight assumes the V shape of the duck. With his glossy back and pure white breast, he is the gayest of our sea birds; and, both sexes being alike, the gaiety is not, as in the case of the ducks, dimmed by the sober-coloured females. He is not an oyster-catcher; if he were, he would probably be persecuted. Your oyster, besides being a bivalve that can be trusted to take care of itself, is much too rare in Scottish waters to provide adequately for the multitude of birds. The mussel is much commoner, and more easily managed. He gains ready entrance by driving a semicircular hole into the shells lying exposed on the banks.

Just beyond the zone of the oyster-catcher, stalking soberly, and every now and then inserting his beak into the sand in search of annelids, is the curlew. What advantage the bend in the bill gives, except perhaps to feel under stones, it were hard to say; and as no other explanation is forthcoming, I shall adopt this one tentatively. The various ways in which the same end is reached always keep the student of nature from wearying. The bill of the snipe is straight; that of the curlew bends decidedly towards him; that of the godwit slightly away from him.

Still farther out than the curlew is a shapeless mass all huddled together, which only experience could identify as the heron. Deep as he is, the short tail is clear above the surface. And, when he rises, the long legs are dropped behind to perform the tail's function as the steering apparatus of the bird. Thus, two useful ends are secured by the simplest possible device.

Round the corner, just within the shelter of the projecting sand-dunes when the wind is easterly and seaward, is an old boat. It drifted ashore one night in a gale; and the owner was either too

lazy, or too indifferent to have it launched again.
And there it lies through summer sun and winter
gale, an object no less picturesque than convenient.
So far from being a wreck, its planks are tight;
its cabin is snug, and provided in addition with
workable fireplace and chimney. Its main use
now is as a shelter to night shooters, who can light
a fire and a candle, boil water for tea, and, over a
pipe, while away, with many a yarn, the time till
the tide makes or the flight begins, in picturesque
comfort.

Should an ugly night come down with no shelter
for a mile or two, it is a veritable haven; and enviable is he who goes to sleep there to the roaring of
wind, the lashing of rain, and the lapping of water.

Its present occupant was a civil fellow, whose
amiable weakness was a preference for sport when
such was available, over tamer and less interesting
ways of making a living. I found nothing worse
about him than that tendency for wandering from
the high-road on to the grass, which, in the higher
ranks, is known as Bohemianism. I find such
people excellent company, with a pleasant dash of
unworldliness about them which affects me as a
green place to the eye; and, moreover, full of curious
and fresh information.

As a rule, to which there are, of course, exceptions, I had rather talk with an intelligent gamekeeper than with the lessee of the shooting, because he is on the spot at all hours of the day, and seasons of the year. And I had rather talk with an intelligent poacher than either, because he keeps irregular hours, and sees wild life when others are asleep. If anyone knew that water, with the nesting-places, and goings and comings of its feathered inhabitants, this man did. In addition to the eye of a naturalist, he had the interest of one to whom exact knowledge meant daily bread.

I mount the boat beside him, and have a look out. Nowhere more than two miles wide, the estuary is at once ample, and yet within manageable limits. Nothing could escape the watcher's glass. But the first thing that appears needs no glass.

A large bird sails into view, and, with a graceful curve, settles on the water. My friend sculls quietly to windward, and prepares for action. According to its wont, the swan rises against the wind, and comes toward him. In a few moments it is on board. It proves to be a mute swan. The man is silent also; the reason being that this species, especially when abroad singly, is usually truant from some ornamental pond, or small lake, where his semi-domesticated kindred are sailing among the water-lilies. There is always the benefit of the doubt, as a flock in a truly wild state occasionally makes its appearance. Another visitor in hard winters is distinguished from "the mute" as "the whooper," because of the noise he makes.

The little punt rose scarcely above the water, and carried a long raking gun, which took in a pound and a half of lead. Altogether it was admirably fitted for its mission as scourge of the estuary.

The *modus operandi* was as follows: A moonlight night was chosen, with a flowing tide. Like the ghost or shadow it was, the punt glided through the water, and lingered opposite the mud flats, or mussel-scaups exposed by the tide. As the flow covered their feeding-ground, the ducks were lifted up, and floated away. When the vast flocks, often of many hundreds, crossed the track of moonlight, the time for action had arrived. The big gun was brought to bear, with disastrous consequences.

It was sheer butchery; not so much because of those that were killed, though that was whole-

sale enough. Many that were winged, or otherwise wounded escaped for the time, only to meet
a worse fate. What that was we shall see
directly. More than once the engineer of all
this mischief was well-nigh "hoist with his own
petard." One night the gun broke loose, pinning
him to the bottom of the punt, and for awhile
he seemed in hopeless plight. When at length he
managed to free himself, he did not stay to collect
the dead.

Of course he was not poaching, since the birds
were not game, but only of the plebeian sort; nor
trespassing, for the waters were not preserved;
nor breaking any human law, for it was not close
time. It was only the man's rough way of
shooting. The best traditions of sport pass down
another channel, and miss him out.

The head and front of his offending was that
he took the full advantage of his liberty, and
went the nearest way to his end, undeterred by
any sentiment. Let us hope that those who
know better never transgress in a similar way;
never seek a big bag, irrespective of the means by
which it is filled; never have drives or battues;
never forget that wild creatures should have
fair play; and that shooting is no longer sport,

and becomes unworthy of a gentleman, from the moment that it ceases to be a contest, on fairly equal terms, between the cunning of the animal and the skill of the man.

There is something to be said for this matter-of-fact puntsman, who transgresses the unwritten code which he never heard of, for the sake of a living; but none for him who makes it matter for silly boast. One may even be permitted to question the soundness of the advice to get a lot of wild fowl in a row before firing; and always to look for a sitting shot as being more certain. Such advice is based on the big-bag theory, and involves, in a lesser degree perhaps, the unpardonable offence charged against the punt gun, of wounding as many as one kills. It seems to put wild fowl outside the pale of ordinary law, as if it were needless to give them the consideration we pay to the winged aristocracy. Why not seek a sitting shot at other birds, first of all being careful to get them in a row? Why not fire into a covey of partridges, delighted if we secure four, where otherwise we should only have got two?

Of course, a good retriever may bring in the severely wounded; but what if it secured every one, however slightly winged; that does not

18

make the practice good form. I can find no valid reason that is not equally applicable all round, why one should not be contented with a right and left even at wild fowl.

Darkening the water for a considerable distance, is an immense flock of scaup-ducks, so named, probably, because they usually feed on the mussel-beds, or scalps. They are winter visitors from the North, and arrive late: should the weather remain open, not till nearly Christmas. During their three months of stay, they are the commonest of our estuary ducks.

There is a prejudice against exclusive feeders on sea organisms as table birds, which ought to shelter many of the ducks, and does shelter some of them from destruction. But, seeing that the scaup-ducks, and many others still more objection-able, find their way to the shops, there must be some means of rendering them palatable. An acquired taste necessarily increases one's range of choice, and, after high game, anything in that direction should be possible.

A large flock of widgeon are scattered over the flat, half a mile farther up. They are day-feeders, but come up at night when the tide is suitable, and so expose themselves to the tender mercies of the big gun. Their curious whistle, so startlingly like that of a man calling a comrade, is uttered as they rise, and during flight; and differs from their cry on the bank, or on the water.

The quacking ducks are the teal, and the mallard. The teal looks a miniature mallard; the mallard a gigantic teal. Both are here: the tiny teal in tiny flocks; the mallard in greater numbers, as becomes his greater size.

While widgeon, and scaup move up and down with the tide, and are seldom absent from the estuary, the mallard has regular habits. One knows exactly when and where to look for him. All day he rides in safety beyond the bar, and takes to the wing just after dusk. Dark comes early at this time of the year. If one takes his stand at the water-side at half-past four, he will not have long to wait. If he is unaware of what is going to happen, he is not unlikely to be startled by the rush of wings.

On calm nights they usually fly high, probably out of sight in the faint light. When there is much wind, they come nearer the ground, and within easier range. The firing of guns records the great speed at which they are travelling. The first sounds near at hand; the second comes faintly out of the distance. A right and left yielding, if successful, a pair, is the most the expertest shot can expect on the passage. More are shot where they alight. The estuary is a guide rather than a goal. They speedily diverge from it, generally where some burn finds its way in, and make for their feeding-ground in inland field or marsh.

Barnacle, brent, and pink-footed geese; scaup and widgeon; mallard and teal; pintail and golden-eye; curlew and oyster-catcher; godwit, knot, and redshank; dunlin and sanderling; silver and ringed plovers; are among the commoner estuary forms in winter.

There are no hawks present to trouble them. The peregrine falcon, both wild and tame,—for this is a historic hawking scene,—struck his quarry here, but now, in neither form, is he any longer present.

One wonders why this picturesque form of sport, so different from the punt gun, even from the right and left, involving a more refined skill, as well as lending itself to the purposes of art, should not be revived, along with other old-fashioned forms. Why the lighter kestrel should no longer sit on my lady's wrist. Our forefathers chose these birds, not blindly, but with a full knowledge of their dispositions and aptitudes; and recent experiments show that they are as docile now as ever.

Not that the estuary is without its birds of prey ; but they are of the common sort. Hooded and carrion crows exist during the winter, in about equal numbers. In the grey morning they can be seen scouting along the shore in search of the dead, and scanning the estuary for the wounded. Already before one can get along, unless he has spent the night in the boat, every bone of every bird cast ashore has been picked. This service ought, perhaps, to be placed to their credit, as it certainly makes the high-water mark sweeter, and the walk pleasanter. But when we have called them scavengers, we are afraid that we have given their only title to honour.

That Golden-eye, one winged in the twilight—if there is any twilight at that season—and hoped to pick up in the morning, has already been detected by sharper eyes than his. The remorseless foe has seen his advantage, and used it.

When disabled prey are not to be had, he will even stalk the capable, and may, occasionally, be seen making the diabolic attempt to drown the water birds in their own element. The great black-backed gull is credited with swallowing the wounded birds without taking the trouble of picking them.

Hard weather greatly increases the variety of estuary life. The green plover—seldom absent— comes down from the frozen fields in larger flocks. The golden plover appears from his devious flight over the Lowlands. The snipe deserts the frozen marshes for the softer mud flats ; and a little later, when the frost has had time to creep in under the shadow of the birches, the woodcock follows from his winter home in the copse. A busy gun can make a rich and varied bag.

Small birds innumerable are driven down—all, perhaps, except the seed birds, which are seldom at a loss ; with the hedge-warbler, the robin, and the ousel, which in their various spheres can manage to pick up a living—until sometimes it is hard to tell what kinds are there ; still harder to tell what kinds are absent.

No scene in nature that I am acquainted with is richer in possibilities of observation, or sport than the winter estuary ; except, perhaps, that twin scene, the winter loch. The two have very much in common. Wild duck retire indifferently to both for their daily siesta ; and rise indifferently from both when twilight summons them to their

feeding - ground. Teal, and widgeon feed indifferently on the grasses which grow round the
margin of both.

The golden-eye, though classed as a sea duck,
inclines to fresh water, and enlivens most of our
inland lochs by his incessant diving.

The pochard also winters in the loch in preference to his native sea. In the sheets which
he frequents, he seems to act as a provider for the
non-diving species. Dropping to the bottom, he
pulls up the water plants, tears off the roots as
his share, and allows the blades, for which he has
no liking, to float on the surface, there to be con-
sumed by the attendant ducks.

A bag from an estuary should include mallard,
teal, and widgeon, with scaup and pintail; a bag
from an inland loch should include mallard, teal,
and widgeon, with pochard and golden-eye. If the
balance of the ducks is thus fairly well preserved,
the estuary has a very great advantage in the
number and variety of the waders.

The inhabitants of the two domains are
frequently interchanging. A gale from the sea
will drive the estuary birds to the land-locked
lakes; a continued frost, by icing over the water,
will send the inland birds from the smaller to the

larger and deeper lochs, and finally to estuaries, for food.

As weeks pass by, a little wave of restlessness trembles over the scene. In the life of birds in general, and estuary birds in particular, are two well-marked annual phases. There is a time to die, and a time to be born. The waste, which must be enormous, they are bound to repair, or perish; and they will soon be off to the quiet moors and marshes; or to the quieter shores beyond the sea, to undertake the mighty task. They are fairly successful. As between life and death, the balance, in most forms, remains slightly in favour of life.

PRINTED BY MORRISON AND GIBB LIMITED, EDINBURGH.

www.ingramcontent.com/pod-product-compliance
Lightning Source LLC
Chambersburg PA
CBHW030342270326
41926CB00009B/929